Walter Ott is Assistant Professor of Philosophy at Virginia Tech.

CAUSATION AND LAWS OF NATURE IN EARLY MODERN PHILOSOPHY

Causation and Laws of Nature in Early Modern Philosophy

WALTER OTT

OXFORD
UNIVERSITY PRESS

OXFORD

UNIVERSITY PRESS

Great Clarendon Street, Oxford OX2 6DP

Oxford University Press is a department of the University of Oxford.
It furthers the University's objective of excellence in research, scholarship,
and education by publishing worldwide in

Oxford New York

Auckland Cape Town Dar es Salaam Hong Kong Karachi
Kuala Lumpur Madrid Melbourne Mexico City Nairobi
New Delhi Shanghai Taipei Toronto

With offices in

Argentina Austria Brazil Chile Czech Republic France Greece
Guatemala Hungary Italy Japan Poland Portugal Singapore
South Korea Switzerland Thailand Turkey Ukraine Vietnam

Oxford is a registered trade mark of Oxford University Press
in the UK and in certain other countries

Published in the United States
by Oxford University Press Inc., New York

First published 2009

British Library Cataloguing in Publication Data

Data available

Library of Congress Cataloging in Publication Data

Ott, Walter R.
Causation and laws of nature in early modern philosophy / Walter Ott.
p. cm.
Includes bibliographical references and index.
ISBN 978–0–19–957043–0 (alk. paper)
1. Causation. 2. Philosophy of nature. 3. Necessity (Philosophy) 4. Natural law. I. Title.
BD541.O83 2009
122.09'032—dc22   2009016665

Typeset by Laserwords Private Limited, Chennai, India
Printed in Great Britain
on acid-free paper by the
MPG Books Group Ltd, Bodmin and King's Lynn

ISBN 978–0–19–957043–0

Acknowledgments

Like anyone working on a big project, I have amassed a number of debts. I hope my list is complete.

My first debt is to Steven Nadler and Donald Rutherford, whose 2004 National Endowment for the Humanities Summer Institute, 'The Intersection of Philosophy, Science, and Theology in the Seventeenth Century,' at the University of Wisconsin–Madison, pointed me in a new direction. I thank the presenters and participants, particularly Michael LeBuffe, Mary Domski, Michael Jacovides, and Martin Stone.

The book owes a great deal to four anonymous referees for Oxford University Press, one of whom provided an unusually detailed set of comments. I thank them for saving me from a number of idiocies. Dan Flage was kind enough to comment on the entire manuscript, and I am greatly indebted to him. Bryan Hall has been an invaluable resource, commenting on my work and sitting in on a seminar I taught on these issues at Virginia Tech in 2005. Michael Jacovides suffered through the early stages of my work on Locke and Boyle on relations and provided indispensable comments. Antonia LoLordo offered helpful criticisms of my work on Malebranche. Leslie MacAvoy helped with my translations of Régis. In correspondence, Tad Schmaltz disabused me of some mistakes. And my deepest thanks to Alan and Ryan Robinson for the cover art.

Excerpts have appeared in different form in the journals: 'Causation, Intentionality, and the Case for Occasionalism,' *Archiv für Geschichte der Philosophie*, 90/2 (2008); 'Régis's Scholastic Mechanism,' *Studies in History and Philosophy of Science*, 39/1 (2008); and 'Hume on Meaning,' *Hume Studies*, 32/2 (2006). I am grateful to the editors of these journals for permission to reprint this material and even more so to the referees who commented on it.

Parts of the book were presented in various venues, and I thank the participants and organizers of the fall 2006 meeting of the South Central Seminar in Early Modern Philosophy, University of Arkansas, Fayetteville; the spring 2007 Pacific Division meeting of the American Philosophical Association, San Francisco; the University of Virginia and Virginia Tech colloquium series, 2007–8; and the 2008 Conference on Causation 1500–2000, York, UK.

Finally, I would like to thank Peter Momtchiloff, Catherine Berry, Tessa Eaton, and the editorial, design, and production staff at Oxford University

Press, UK. I am particularly grateful to Laurien Berkeley, the Press's copy-editor, for important refinements and corrections throughout.

W.O.

Charlottesville, Virginia
March 15, 2009

Contents

PART III: POWER AND NECESSITY

PART IV: HUME

Abbreviations and Conventions

AG   Gottfried Wilhelm Leibniz, *Philosophical Essays*, ed. R. Ariew and D. Garber (Indianapolis: Hackett, 1989)

AT   René Descartes, *Œuvres de Descartes*, ed. C. Adam and P. Tannery, 11 vols (Paris: J. Vrin, 1904)

CPA   St Thomas Aquinas, *Commentary on the Posterior Analytics*, trans. F. R. Larcher (New York: Magi Books, 1970). Cited by book, lecture, page

CQ   Robert Boyle, 'Of the Systematical or Cosmical Qualities of Things' (1670), in *The Works of Robert Boyle*, ed. M. Hunter and E. Davis, 14 vols (London: Pickering and Chatto, 1999), vi

CS   Robert Boyle, 'Cosmical Suspicions' (1670), in *The Works of Robert Boyle*, ed. M. Hunter and E. Davis, 14 vols (London: Pickering and Chatto, 1999), vi

CSM   *The Philosophical Writings of Descartes*, 3 vols (New York: Cambridge University Press, 1984), i and ii, ed. John Cottingham, Robert Stoothoff, and Dugald Murdoch

CSMK   *The Philosophical Writings of Descartes*, 3 vols (New York: Cambridge University Press, 1984), iii, ed. John Cottingham, Robert Stoothoff, Dugald Murdoch, and Anthony Kenny

DM   George Berkeley, *De Motu* (1721), in *The Works of George Berkeley*, ed. A. A. Luce and T. E. Jessop, 8 vols (London: Thomas Nelson, 1949–58). Cited by section

E   David Hume, *An Enquiry Concerning Human Understanding* (1748), ed. T. L. Beauchamp (Oxford: Clarendon Press, 2006). Cited by section and paragraph

EMH   Robert Boyle, 'About the Excellency and Grounds of the Mechanical Hypothesis' (1674), in *Selected Philosophical Papers of Robert Boyle*, ed. M. A. Stewart (Indianapolis: Hackett, 1991)

Essay   John Locke, *An Essay Concerning Human Understanding* (1690), ed. P. H. Nidditch (Oxford: Clarendon Press, 1975). Cited by book, chapter, section, page

IPQ   Robert Boyle, 'An Introduction to the History of Particular Qualities' (1671), in *Selected Philosophical Papers of Robert Boyle*, ed. M. A. Stewart (Indianapolis: Hackett, 1991)

MD   Francisco Suárez, *Metaphysical Disputations* (1597), in Suárez, *On Efficient Causation*, trans. A. J. Freddoso (New Haven: Yale University Press, 1994). Cited by question, section, page

NN   Robert Boyle, 'A Free Inquiry into the Vulgarly Received Notion of Nature' (1686), in *Selected Philosophical Papers of Robert Boyle*, ed. M. A. Stewart (Indianapolis: Hackett, 1991)

OC   Nicolas Malebranche, *Œuvres complétes de Malebranche*, 20 vols (Paris: Vrin, 1958–84)

OFQ   Robert Boyle, 'The Origin of Forms and Qualities according to the Corpuscular Philosophy' (1666), in *Selected Philosophical Papers of Robert Boyle*, ed. M. A. Stewart (Indianapolis: Hackett, 1991)

PHK   George Berkeley, *A Treatise Concerning the Principles of Human Knowledge* (1710), in *The Works of George Berkeley*, ed. A. A. Luce and T. E. Jessop, 8 vols (London: Thomas Nelson, 1949–58). Cited by part or introduction and section

RD   Robert Boyle, 'An Essay, Containing a Requisite Digression, Concerning those that would Exclude the Deity from Intermeddling with Matter' (1663), in *Selected Philosophical Papers of Robert Boyle*, ed. M. A. Stewart (Indianapolis: Hackett, 1991)

SAT   Nicolas Malebranche, *The Search After Truth* (1674–5), trans. T. M. Lennon and P. J. Olscamp (Cambridge: Cambridge University Press, 1997). Cited by book, chapter, section, page

*SAT*E   Nicolas Malebranche, 'Elucidations of *The Search After Truth*' (1678–1712), in Malebranche, *The Search After Truth*, trans. T. M. Lennon and P. J. Olscamp (Cambridge: Cambridge University Press, 1997). Cited by Elucidation and page

SCG   St Thomas Aquinas, *Summa Contra Gentiles*, in Aquinas, *Basic Writings*, ed. Anton Pegis, 2 vols (New York: Random House, 1945), ii. Cited by chapter and page

SGP   Pierre-Sylvain Régis, *Cours entier de philosophie, ou, Système général selon les principes de Descartes* (Paris, 1690)

ST   St Thomas Aquinas, *Summa Theologica*, in Aquinas, *Basic Writings*,
          2 vols, ed. Anton Pegis (New York: Random House, 1945), i. Cited
          by part, question, article

T   David Hume, *A Treatise of Human Nature* (1739–40), ed. D. F.
          and M. Norton (Oxford: Clarendon Press, 2000). Cited by
          book, part, section, paragraph; or introduction and paragraph;
          or Abstract/Appendix page number

URF   Pierre-Sylvain Régis, *L'Usage de la raison et de la foi* (1704), ed.
          J.-R. Armogathe (Paris: Fayard, 1996)

Introduction

This book takes up a handful of closely related lines of argument concerning causal powers and laws of nature and pursues them through the works of some of the most important figures of the early modern period. I have not attempted a definitive history of these controversies, nor have I aimed to give a brief overview of them. The former project is daunting, to say the least, while the latter has already been accomplished.[1] Instead, I have selected a manageably narrow set of themes and figures in the hope that I will be able to shed light, not only on the contours and content of the historical debate, but also on the philosophical issues themselves.

In the hundred or so years between the publication of Descartes's *Meditations* and Hume's *Treatise*, something like our contemporary notion of a law of nature is developed and realism about causation is challenged.[2] This is no accident; the one development made the other possible, if not inevitable.

Hume famously calls causation 'the cement of the universe,' and in the early modern period, that cement is gradually dissolved. It would be myopic to blame (or credit) Hume alone for this. The stage was set by Descartes's invention of laws of nature. Prior to Descartes, expressions like *lex naturalis* had a use primarily in the context of divine command theory.[3] A law of nature was a normative claim that had its source in the divine will. When Descartes transfers this notion to the realm of physical objects, it retains some of its original features. In particular, it still implies that the law in question is arbitrary, since it is founded on the will of the lawgiver.[4] This conception exercised a powerful influence, not only over Cartesians like Nicolas Malebranche, but also over George Berkeley, and in some

---

[1] See Kenneth Clatterbaugh (1999).

[2] A note on 'realism': this is a notoriously slippery term. All of the figures we shall examine are causal realists in the sense that they believe there are such things as causes. Even Hume, of course, believes this. So by 'causal realism' I shall mean, not just a view that acquiesces in causal talk, or holds that causal statements have a truth value, but the view that there is a mind-independent connection between the distinct objects or events that qualify as 'cause' and 'effect.'

[3] This is not to say that some authors did not at times use the concept of law as a metaphor by which to express God's immutability or the regularity of natural events. See below, Ch. 7.

[4] This does not, of course, mean that the law is arbitrary *tout court*, since the will of the lawgiver might be constrained in some way. Indeed, a typical position is that God's nature constrains, if not determines, his choice of law. So although calling such laws 'arbitrary' is hyperbolic, it brings out the crucial point that these laws are not determined by the nature of the objects that 'obey' them.

moods, Robert Boyle. On this view or family of views, God is the immediate source of the laws of nature and hence the primary, if not the only true, cause. For objects can be said to 'obey' these laws only in an attenuated sense. If God is to give the army of unalterable law its marching orders, he must move his soldiers about himself.

On the older view, which Descartes and his fellow travelers sought to supplant, the core notion of natural philosophy is not law but power. The powers of created beings are at least partly responsible for body–body interactions and help determine the course of events. Although Descartes et al. mount a withering attack on this Aristotelian position, they hardly succeed in pulling it out by the roots. Thus, in the same period we find an opposing family of views that attempts to meld the Aristotelian's commitment to the causal powers of bodies with the new ontology of mechanism. On this picture, God does not directly will how bodies are to move; instead, bodies have, to some degree or other, genuine causal powers that explain why they do what they do. In this vein, we find such disparate authors as Pierre-Sylvain Régis, Boyle (in other moods), and John Locke struggling to remake Aristotelian powers in the mechanist image.

The debate between these two factions is the central theme of this book. It follows the pitched battle over laws and powers to its conclusion: Hume's wholesale rejection of causal realism. Abandoning powers for many of the same reasons as his predecessors, and constrained by a picture of intentionality that makes a realist position unintelligible, Hume ends up seeing all events as 'entirely loose and separate,' connected neither by their own natures nor by the will of a creator. This is of course tied up with a number of other issues in the period I shall discuss, including the ontology of relations, necessity, and meaning.

Although I shall impose a narrative on the debate, I am wary of taking such heuristics too seriously. As the guarded talk of 'inevitability' above signals, I do think some positions are the logical offspring of others. More controversially, perhaps, I think some positions are possible only given a prior context of thought. Some contemporary conceptions are simply not available, I shall argue, in the early modern period. At the same time, I would not want to be construed as suggesting that the modern period represents a uniform series of improvements on more or less primitive views. Some moves in the debate strike me as decided steps backward. For example, I am far from holding that Humean anti-realism is the logical consequence of Lockean empiricism. As I shall argue, Hume's rejection of definite descriptions as a means of referring to causal powers is ill-founded, and as a consequence, some of his arguments against causal realism are impotent. If I am right about such mis-steps, then, in the words of Mark Borchardt, 'the inevitable has become quite the evitable.'[5] What my story lacks in Hegelian drama will, I hope, be made up for in accuracy.

---

[5] *American Movie* (1999).

To get at the central debate without bogging down in endless details, I have chosen largely to ignore a number of connected issues that are interesting enough in their own right.

First, I shall be mainly interested in causal interactions between physical objects. While mind–body and body–mind causation are of the first importance for metaphysics, they have been widely covered in other works. At times, of course, a philosopher's treatment of these other causal interactions will be used to illuminate his treatment of body–body causation.

Second, there are a number of issues about causation that do not directly connect with my interest in the relationship between causal powers and laws. Whether a cause must always be prior to its effect, for example, exercised many of the philosophers I shall discuss. Given my purposes, it is of little importance. Similarly, the gradual contraction of Aristotle's four causes into one, namely, the efficient cause, will go largely undiscussed, partly because, as with the rationalists' development of a strong principle of sufficient reason, it has been well and thoroughly covered elsewhere.[6]

Third, I shall focus on a limited number of figures, not all of whom have been fully accepted by the canon. And some major figures, such as Spinoza and Leibniz, will at most have cameos. Their views on causation have already been subjected to a fair bit of scrutiny.[7] It then makes sense to devote a bit more time to, say, Locke, who is still too often regarded as a halfway house on the road to Hume. But the real reason for the neglect of Spinoza and Leibniz is their minimal influence on the debate in England. Apart from Leibniz's controversy with Newton, they are marginal figures. And Spinoza seems to have been known mainly by way of Bayle; his name becomes a term of abuse for any departure from religious orthodoxy. Since my goal is to dissect a web of argument extending from Descartes to Hume, Spinoza and Leibniz must lurk in the background. By contrast, there are direct connections linking the figures I have chosen. Berkeley and Hume, for instance, were both directly influenced by Malebranche, Hume to the point of all but plagiarizing from his copy of Thomas Taylor's translation of *The Search After Truth*.[8]

Finally, my interest is squarely in the metaphysics of causation and law. Although epistemic issues, particularly induction and explanation, will feature at various points in the project, they are peripheral. And although the scientific debates of the age will obviously have a bearing on what follows, they are not

---

[6] For an exhaustive treatment of these themes in Suárez, Descartes, Malebranche, Spinoza, and Leibniz, see Carraud (2002). As the title of his book indicates, Carraud's theme is the connection between causation and intelligibility, particularly as enshrined in the principle of sufficient reason. Carraud thus devotes a good deal of time to Descartes's doctrine of the divine creation of eternal truths and rationalist responses to it. These issues are orthogonal to my main project.

[7] For example, Steven Nadler's edited volume *Causation in Early Modern Philosophy* (1993*a*) contains essays on these figures and other rationalists, but not a single essay on an empiricist. Clatterbaugh (1999) is similarly thin on the empiricists.   [8] See McCracken (1983).

my central concern. Particular issues, such as those of cohesion and gravity, will sometimes come to the fore, especially in Part III. But again, my focus throughout is on the underlying metaphysical issues.

I have tried to write a book one might enjoy reading. To that end, many of my disagreements with other commentators have been relegated to footnotes. I mention them in the text only when they are useful foils in presenting my own view. Inevitably, some portions of the book are, as one reader gently put it, 'tough sledding.' This is particularly true of Chapters 6, 8, and 9. The reader whose main interest is not in Descartes can profitably read the rest of the book while skipping over these sections, with the possible exception of section 9.2.

1

Themes

1.1 THE ORIGIN AND STATUS OF LAWS OF NATURE

Although Descartes speaks in terms of 'laws of nature,' this is something of a misnomer, since the foundation of those laws is not nature in one of its usual senses—the ordinary physical world—but rather God and God's will. In the *Principles of Philosophy*, Descartes announces that 'From God's immutability we can...know certain laws or rules of nature, which are the secondary and particular causes of the various motions we see in particular bodies' (AT viiiA. 62/CSM i. 240). All three of Descartes's laws are deducible from God's unchangeability. That an object in motion continues to move in a straight line unless impeded, for example, is a result solely of God's uniform behavior.

Now, as Boyle later points out, there is something odd in Descartes's notion of a law. For laws in the sense specified by divine command theory can apply only to beings who are capable of reasoning. '[T]o speak properly, a *law* being but a *notional rule of acting according to the declared will of a superior*, it is plain that nothing but an intellectual being can be properly capable of receiving and acting by a *law*.'[1] Boyle thus regards Descartes's usage as nothing but a deeply misleading metaphor.[2] Since material substance's properties are exhausted by those of its essence, extension, it clearly cannot be said to obey laws in the way a prince's subjects obey him. Descartes seems to mean, then, that matter obeys these laws only in the sense that God moves bits of matter around in accordance with them. I argue for this in detail below; it is a controversial claim. What is important at the moment is that these laws float free of the matter whose behavior they prescribe. To capture this feature of Descartes's view, we might call his a 'top-down' analysis of the laws of nature. These laws are not fixed by the natures of the objects they

---

[1] NN 181. Boyle might have been responding here to Richard Hooker (1593), among others. Hooker writes as if inanimate objects could literally be said to obey a law. He writes, 'Whereas therefore things naturall which are not in the number of voluntarie agents, (for of such only we now speake, and no other) do so necessarily observe their certain lawse, that as long as they keepe those formes which give them their being, they cannot possiblie be apt or inclinable to do otherwise than they do' (quoted in Milton 1998: 681).

[2] At least in some texts. As I argue below (Ch. 17), Boyle occasionally says quite the opposite, and seems to endorse a kind of Cartesian voluntarism.

govern; both their status and their content depend not on created beings but on God.

Contrast a 'bottom-up' view, which holds that the course of nature is fixed by the properties of created beings. On this position, fire's tendency to burn dry wood under standard conditions is a function of the powers of the fire and wood; to create a world in which all of the conditions are right and yet the wood remains unsinged, God would have to create a world in which neither fire nor wood exist.[3] Once the relevant properties are instantiated, nature takes the course it does simply in virtue of the kinds of things that make it up. On this view, the properties of objects, not the will of God, play the fundamental role. 'Laws of nature' will then be nothing more than convenient ways to state relations among these properties.[4] Though the distinction between top-down and bottom-up views is neither so rigid nor so sharp as I here imply, it is a helpful beacon in heavy seas.

I shall say more below about the origin of these concepts of law, and their relation to earlier, scholastic conceptions.[5] For now, it is worth turning in the opposite chronological direction and noting that neither of these concepts maps on very well to at least one common contemporary use of 'law.' If I had to describe the default view held by the average philosopher, my (former) self included, I would say something like this. Laws of nature are sets of truths that define a region of modal space: the nomologically possible. This region includes all that can happen, given those laws. Outside this modal space lies the logically possible, which, although it of course includes the nomologically possible, includes much that is logically consistent but in violation of the laws of nature of our home world. Most of us find it easy, for example, to conceive of possible worlds in which the same objects exist, with the same properties as in our own, and yet the

---

[3] But see my treatment of miracles in the concurrentist tradition (below, sect. 3.1).

[4] It is worth noting that some philosophers use 'law' as replacement for 'form' or 'quality.' For example, Bacon writes that 'though nothing exists in nature except individual bodies which exhibit pure individual acts in accordance with law, in philosophical doctrine, that law itself, and the investigation, discovery and explanation of it, are taken as the foundation both of knowing and doing. It is this *law* and its *clauses* which we understand by the term Forms' (*New Organon*, in Bacon 2000: 103; see 119, 127, 131). Similarly, Newton says that he takes principles like cohesion and gravity not as 'occult qualities, supposed to result from the specific forms of things, but as general laws of nature, by which the things themselves are formed' (2004: 137).

[5] See Ch. 7 below. Steinle (2002) does an excellent job of cataloging the widely disparate uses of 'law' throughout the modern period, particularly in the Royal Society. Even in Newton's works, Steinle finds that ' "Law" could well stand for the central, axiomatic principles of a field, as well as for more or less empirically established regularities, or for anything in between' (2002: 206). Steinle persuasively argues that the wide appeal of law-talk, and its quite various applications throughout the modern period, is due in part to its suggestion of a divine lawgiver, which affords a prima facie defense against charges of incipient atheism. I shall not rehearse Steinle's arguments or taxonomy here, but the differences among conceptions on the spectrum Steinle speaks of (which, I should emphasize, is quite different from the top-down–bottom-up spectrum) will become important later. See the discussion of analyses (as opposed to models) of law below.

laws of nature are different, leading events involving these objects and properties to turn out equally differently.[6] This conception is utterly foreign to the period we are about to investigate. It clearly has something in common with top-down views, and yet obviously it does not appeal to God as the origin of law or as a continual and ubiquitous source of motion. This contemporary notion of a law is detached from any explanatory base, whether in the natures of bodies themselves or in the activity of a supernatural agent. Perhaps laws really are brute facts about the world; but this is not a view that would have made sense to any of the figures I shall discuss.[7]

Nor is Newton an exception here. It would be easy to think that his predecessors in natural philosophy were simply backward, and something like the conception of law sketched above had to await Newton. While it may be true that Newton is the source of the view I have described, he is, I shall argue, an unwitting one. For he remains (officially) agnostic on the source of the laws he speaks of; this is quite different from denying that they *have* a source.

First, we need to see that the concept of force Newton deploys in the *Principia* is 'mathematical' as opposed to 'physical.'[8] In Definition 8, Newton announces that he proposes to consider forces such as attraction, impulse, and propensity generally 'not from a physical but only from a mathematical point of view,' since he is 'not now considering the physical causes and sites of forces.'[9] That is, Newton's goal is to provide a quantitative treatment of forces while remaining entirely neutral on the ontological underpinnings of the forces themselves.[10] This mathematical approach allows Newton to defend himself against Leibniz's claim

---

[6] I believe this conception to be widespread among non-specialists. It can be found at work in many of the examples and counter-examples offered in epistemology and metaphysics.

[7] Realist contemporary views can be ranged along the top-down–bottom-up spectrum, with the caveat that contemporary views that I put below in the 'top-down' category omit God or any supernatural agent. These views are 'top-down' only in the sense that they do not take laws to supervene on the natures and properties of objects. Now, at the extreme of the top-down conceptions, we might put John Carroll's non-reductive view, which takes laws of nature to be *sui generis* truths that govern all possible worlds. For Carroll, these laws do not flow from the natures of bodies, their properties, or even the regularities exhibited in a given world. Instead, they fix or explain (some or all of) these regularities. At the next point on the scale we can place the Dretske–Tooley–Armstrong view, which analyzes laws as relations of necessitation between universals. On this view, the contingent relation N between F-ness and G-ness makes it necessary that an instance of F-ness be followed (or accompanied) by an instance of G-ness. At the bottom-up end of the scale, we find Brian Ellis's view that laws are reducible to relations between the powers or dispositions of bodies. This is very like the Aristotelian view, since, for Ellis, these powers are not themselves reducible to anything more fundamental. Further along, we might put Nancy Cartwright's (1980) view, with its rejection of the 'facticity' of laws and her emphasis on causal powers. Finally, off this scale altogether, we find the anti-realist 'Humean' view, associated with David Lewis, which, with increasing sophistication, has attempted to reduce laws to regularities among events.   [8] For more on Newton's method, see esp. Smith (2002) and Stein (1990).

[9] In Newton (2004: 64, 63).

[10] See also *Principia*, Scholium to book I, sect. 11, in Newton (2004: 86), and Newton's anonymously published *Account of the Book Entitled 'Commercium Epistolicum'* (2004: 124).

that gravity is an occult quality.[11] The *causes* of gravity might well be said to be occult, as it cannot be explained mechanically; but gravity itself is a perfectly real, fully observable 'manifest quality.'[12] And, as Newton puts it, 'It is not the business of experimental philosophy to teach the causes of things any further than they can be proved by experiments.'[13]

When we try to extract from Newton's texts an account of the origin of gravity, it becomes clear that Newton would reject the most common contemporary picture of laws of nature.[14] This comes out most clearly in his correspondence with Bentley, where Newton famously argues that 'it is inconceivable that inanimate, brute matter should, without the mediation of something else, which is not material, operate upon and affect other matter without mutual contact.'[15] 'Gravity must be caused by an agent acting constantly according to certain laws; but whether this agent be material or immaterial, I have left to the consideration of your readers.'[16] On the view suggested here, gravity is not itself a law but rather the result of a being operating on bodies *according to* a law. Now, to say that the responsible agent operates 'according to certain laws' might mean that it formulates to itself certain conditional propositions and then obeys them, or it might merely mean that the agent operates in regular ways. There is no way to know which picture Newton had in mind; both, I shall argue, appear in other texts in the modern period, particularly in Malebranche's work. What this passage does seem to rule out, however, is the contemporary assumption that laws are autonomous, brute features of the universe: instead, they are the rules by which an unknown agent works. This is true whether those rules are merely summaries of the behavior of that agent or propositions by which it regulates its activities.

One begins to suspect that Newton's rejection of the project of finding the hidden cause of the laws of nature is gradually transformed from a merely epistemic posture to an ontological one. Seeing how well Newton got on without talking about the bases of the laws of motion or of forces, natural philosophers became gradually accustomed to thinking of those laws as autonomous features of the universe that do not stand in need of any such basis.[17]

---

[11] For Leibniz's attack, see esp. 'Against Barbaric Physics: Toward a Philosophy of What There Actually Is and Against a Revival of the Qualities of the Scholastics and Chimerical Intelligences' (AG 312–20). For more on the charge of occultism, see below, sect. 3.3.

[12] In Query 31 to the *Opticks*, Newton writes, 'these [principles, such as gravity] are manifest qualities, and their causes only are occult' (Newton 2004: 137). See also Newton's unpublished letter of 1712, a reply to Leibniz (Newton 2004: 116).   [13] *Account*, in Newton (2004: 123).

[14] By the 'common picture' I mean the view most widely held by philosophers generally, not necessarily by those philosophers of science and metaphysicians who focus on the problem of laws.   [15] Feb. 25, 1692/3, in Newton (2004: 102).

[16] (2004: 103).

[17] This is admittedly speculative; what becomes of the notion of law after the period I am discussing will not form part of my project in this book.

However that may be, Newton's comments offer us an opportunity to clear up a potentially serious confusion. In contrast to the models of laws—top-down and bottom-up—that are the focus of this section, let us call an 'analysis' of laws a statement of their logical form. While Newton's claims above commit him to a top-down model, they are compatible with two kinds of analyses: a conditional analysis, whereby laws are statements of conditionals, and a summary analysis, where laws are mere summaries of regularities. It is tempting to think that the decision between these two analyses will commit one to either a top-down or a bottom-up model of laws. In fact, they are quite neutral with regard to the competing models.

Consider first the conditional analysis: a law of nature states that, if certain antecedent conditions are met, a given effect will follow. This picture is certainly typical of top-down views, such as those of Descartes and Malebranche. Once we push a bit deeper, though, we find that it is not necessarily part of a top-down model.

To see this, consider that we can always ask, in virtue of *what* do these conditionals obtain?[18] One alternative is to insist that they are brute features of the universe. This is not a live option for the figures we shall examine. Another answer, that of Descartes and Malebranche, is that they hold only in virtue of God's volitional activity. That is, God wills that an object or event[19] of type F in circumstances C will be followed by an object or event of type G. But this conditional itself, I shall argue, obtains only in virtue of God's actually willing, on each occasion that Fa and C obtain, that Gb will follow. On this view, laws supervene on God's ubiquitous activity. I shall argue that the notion that the laws of nature 'enforce themselves,' or operate independently of God's individual acts of willing, is just as foreign to these thinkers as the 'brute fact' (non-)analysis. For now, the key point is that the conditional analysis can accommodate any view that grounds out laws in terms of God's activity.

At the same time, partisans of the bottom-up picture can equally well help themselves to the conditional account. Again, in virtue of what do these conditionals obtain? Nothing in the conditional account per se rules out the possibility that the logical grounds of these conditionals are the powers of the individual bodies mentioned in those conditionals.[20] Precisely parallel considerations apply to what I call the 'summary' analysis, where laws are just

---

[18] Cf. George Molnar's application of the *Euthyphro* dilemma to competing analyses of causation (2003: 187).

[19] The moderns in general do not carefully distinguish between objects and events in causal contexts, and I shall be similarly free in my use of these terms.

[20] Even the Humean anti-realist can of course deploy a conditional analysis of law statements. This is the sort of thing envisioned for example by David Lewis and Barry Loewer, according to which a statement counts as a law just in case it is a theorem of the best (simplest and strongest) deductive system (see Loewer 2004: 181 ff.)

generalizations over regularities. These regularities might be grounded in God's activity, the joint activity of God and creatures, the powers of bodies, or, as Hume has it, nothing at all.

What all of this means is that merely deciding whether a figure goes in for a conditional or summary analysis *by itself* cannot tell us where he falls on the top-down–bottom-up continuum. The real question is not which analysis of the logical form of lawlike statements is right, but rather what it is in virtue of which these statements hold. And this is something that can only be captured by the distinction between *models* of laws I have introduced.

With the distinction between top-down and bottom-up models in mind, we can usefully add another pair of categories, beyond those of rationalism and empiricism, with which to carve up the territory of modern philosophy. Some philosophers hold that mechanical explanations stand on their own two feet; for them, the course of events in the natural world is fixed by the nature of that world. In this category we might put Régis, Hobbes, Boyle in some moods, and, I shall argue, Locke. On the other hand, some take such explanations to be incomplete at best and turn either to the divine will, or nowhere at all, for an explanation of the course of events. In this category, we find, surprisingly enough, Descartes and Hume, among others. This metaphysical distinction, based on opposed attitudes toward the causal and explanatory powers of bodies, might in the end prove more illuminating than the purely epistemological contrast embodied in 'rationalism' and 'empiricism.'[21]

The divergence between top-down and bottom-up views will be central to the first three parts of the book. One of the more obscure figures I discuss, Régis, is important precisely because he tries to carve out a way for created beings to contribute to the course of natural events while preserving the Cartesian insistence on God as the sole cause of motion in body–body interaction. What is more, Régis does so by adopting features of scholastic concurrentism for which most other moderns have nothing but scorn. And it is precisely this desire to bring scientific explanations down to earth that leads to the resurrection of the Aristotelian notion of powers. Boyle and Locke take matters even further by paring down, to varying degrees, God's role, and relying even more heavily on powers. But to do so, they had first to devise a notion of power compatible with mechanism.

1.2 THE ONTOLOGY OF POWERS

With the notable exception of Leibniz, the moderns generally regard scholasticism as a tangled skein of outdated and hopelessly obscure doctrines. Locke speaks of

---

[21] Ayers (1996) makes a broadly similar suggestion.

the 'learned Gibberish' of the schoolmen, while Joseph Glanvill, in a memorable phrase, calls Aristotelian science a 'flatulent vacuity.'[22] One of the most frequently ridiculed scholastic notions is that of power.

On the Aristotelian view, a thing's powers derive from its substantial form.[23] Natural, as opposed to artificial, entities have a genuine unity as objects in virtue of their form; this form grounds a further unity among objects that share it. What makes something the kind of thing it is and what that thing can do are equally functions of its form. Although couched in recondite terms, the doctrine, put thus baldly, seems commonsensical enough.

But explanation in terms of powers courts tautology. As Glanvill writes, to say that fire burns by virtue of its heat 'is an empty dry return to the Question' and 'no better account than we might expect of a *Rustick*.'[24] And everyone knows Molière's joke about opium working in virtue of its 'dormitive power.' Thus, explanations in terms of powers came to seem vacuous and circular.

This is not to say that power attributions are *predictively* vacuous. Consider what follows from our dormitive power attribution: claims such as, 'if I take another one, I'll pass out'; 'this will make little Bobby stop coughing.'[25] Such attributions are not entirely vacuous even as explanations: 'when I fell asleep last night, it was because I took that pill and not because I turned on Charlie Rose.' These are precisely the sorts of things no good Humean could accept, at least not without some energetic paraphrasing. The real question, I shall argue, is whether power-based explanations and predictions rest on some more fundamental fact. While it is useful to know that a pill has a dormitive power, it is hard to believe that its disposition to cause sleep is an unanalyzable fact. Intuitively, the same seems to hold for dispositions generally: these need a categorical base.[26]

Thus, a further source of discomfort with powers, and hence with a robust bottom-up position, is their awkward ontological status. If powers are irreducible features of the world, they swing free of the categorical properties of the bodies that have them. And Descartes and Malebranche argue that attributing powers to objects is sheer anthropomorphizing, since powers, as the scholastics think of them, must have characteristics only attributable to minds, such as intentionality. Powers have what the medievals call *esse-ad*, or 'being-toward.' That is, powers are directed at those states of affairs in which they are or would be actualized, even if, as it happens, those states of affairs never obtain.[27]

---

[22] Locke, *Essay*, III. x. 9: 495, and Glanvill (1665: 143).

[23] This is intended only as a first approximation of one common scholastic view. For a more detailed treatment, see Ch. 3 below. A note on terminology: when speaking of the scholastics, I take a thing's *potentiae* to include all of its active and passive powers, each of which flows from its nature; by *vis*, I mean the actualization of an active power.   [24] Glanvill (1665: 126).

[25] Nelson Goodman (1983) and Brian Ellis (2002) make similar points.

[26] Against this, see Ellis (2002, ch. 4).

[27] For a contemporary treatment of powers that accords them 'physical intentionality,' see Molnar (2003); for a treatment of the Aristotelian view in terms of physical intentionality, see

In their effort to carve out a place for the causal contributions of created beings, Locke and Boyle attempt to resurrect a notion of power and thus turn Descartes's top-heavy view upside down. In order to do so, however, they must recast powers in terms acceptable to the mechanist ontology. Having rejected intrinsically intentional physical states as firmly as Descartes, they must find a place for powers among what Boyle calls 'the catholic affections' of matter: roughly, size, shape, and movement. I shall argue that by treating powers as relations, and then reducing relations to the bases on which they supervene, Boyle and Locke attempt to effect this 'sanitizing' of the notion of power. By doing so, they hope to defuse not only the ontological objection but the circularity objection as well, since an effective reduction of the notion of power turns it into a promissory note for an explanation in terms of the primary qualities of matter.

Their work in this direction is not only interesting in itself but provides the most immediate target for Hume's attacks.[28] The road to a mechanically acceptable notion of power was a far bumpier one than these brief remarks suggest, however, and we shall see Boyle and perhaps even Locke failing to carry through with the reduction.

1.3 NECESSITY

Consider Hume's quick argument, itself drawn from Malebranche, against the claim that any two events can be causally connected:

Now nothing is more evident, than that the human mind cannot form such an idea of two objects, as to conceive any connexion betwixt them, or comprehend distinctly that power or efficacy by which they are united. Such a connexion wou'd amount to a demonstration, and wou'd imply the absolute impossibility for the one object not to follow, or to be conceived not to follow upon the other: Which kind of connexion has already been rejected in all cases.   (*T* 1. 3. 14. 13)[29]

The real puzzle about this kind of argument has never been its form or structure but rather who is supposed to be bothered by it. G. E. M. Anscombe (among others) has suggested that Hume's target is the conception of causation as logical necessitation.[30] But even this seems insufficient, since Hume takes conceivability to be a test of possibility, and if a logical truth is sufficiently complicated, our

---

Des Chene (2006). A helpful discussion of the intentionality of powers in Aristotle and Aquinas can be found in Adams (2007).

[28] I mean 'immediate' in the sense of temporal proximity; as I shall argue, Hume is equally concerned to defeat Aristotelian views.

[29] Following Steven Nadler (1996), we can call this the 'no necessary connection' argument, or 'NNC' for short.   [30] Anscombe (1993).

powers of conceiving might be insufficient to tell us whether or not what we are conceiving is even logically possible. Even if we put this aside, it is not easy to imagine someone arguing that causal claims can be known a priori. That fire causes heat is obvious enough in light of our experience; but how could anyone think that our ability to conceive of a state of affairs in which fire does not produce heat shows anything at all about the causal claim? Why would anyone think that causal relations are governed by logical necessity? If Hume merely defeats *this* view, however roundly, we should be underwhelmed.

It is increasingly common to read that Hume's arguments cannot touch a realist view that relies on *nomo*logical necessity. Writing in 1990, Nicholas Jolley noted that 'Today it is natural to object that while genuine causal connections are indeed necessary, the necessity in question is not logical.'[31] Thus, when Hume uses this argument to show that there can be no necessary, and hence no causal, connection between any distinct objects on the grounds that he can conceive of those objects existing independently, he is conflating causal and logical necessity. And according to some commentators, Hume's neglect of nomological necessity is doubly unforgivable, since such necessity was a key part of traditional Aristotelianism, just as it is in the contemporary 'Aristotelianism' of Ruth Barcan Marcus et al.

For my part, I think the claim that modern and pre-modern Aristotelianism has at its core a commitment to nomological necessity is deeply wrong-headed. I shall argue below (Chapter 3) that the typical Aristotelian position holds that sublunary events are linked by what we would call logical necessity: it is logically impossible, and hence inconceivable, that a cause not produce its effect, under standard conditions. What is more, this view pops up in some surprising places throughout the early modern period. It lurks behind Locke's vision of a demonstrative science of the natural world, where we can, in principle, 'know without Trial' how bodies will operate on one another (*Essay*, IV. iii. 25: 556); and it helps to explain Malebranche's otherwise unmotivated claim that 'A true cause . . . is one such that the mind perceives a necessary connection between it and its effect' (*SAT* VI. iii. 2: 450), a claim that plays a vital role in both his and Hume's arguments.[32]

If my view is right, the easy reply to what has come to be known as the 'no necessary connection' argument is simply out of place, since that argument is not directed against the doctrine of nomological necessity. This, of course, lets

---

[31] Jolley (1990*a*).

[32] I should note that I am using 'logical necessity' to mean simply truth in every possible world, and 'logical possibility', truth in at least one such world. (This is intended as an elucidation, not as a definition; otherwise, it would be viciously circular.) I do not mean to suggest that all logical impossibilities are self-contradictory, or, correspondingly, that all logically necessary truths are analytic. Locke, for example, takes the truths of geometry to be at once synthetic and logically necessary.

us make more sense of Hume's (and Malebranche's) use of the argument: they are neither so helplessly ignorant of their antagonists' views, nor so quick to set up straw men, as some contemporary commentators suppose.

1.4 MODELS OF CAUSATION

I shall be arguing that causal realism in the modern period begins with the Aristotelian conception of causation as logical necessitation and never really breaks free of it. This is not to say that the realists among the figures I discuss speak with one voice where the nature of such necessitation is concerned. We can discern two quite different models of causal connection in the period, both of which have roots in the scholastics, though they grow in very different directions. While these models are most clearly articulated by Malebranche and Locke, respectively, they exercise a wide influence on later philosophers and have correspondingly deep roots in earlier ones.

The first model, which I call the 'cognitive' model, springs from the *esse-ad* feature of powers. Aristotelian powers, as we have seen, are directed at non-actual states of affairs; this leads Descartes to mock them as 'little souls,' capable, like minds, of intentionality. On the scholastic view, the tie between the exercise of a power and its effect lies in the nature of that power itself, which includes its effects in the sense that it is defined by them. Though it sounds odd to contemporary ears, truths about powers are analytic truths that can be discovered only through repeated experience. Any true causal claim, then, will be logically necessary: a world in which fire fails to burn paper is a world that lacks either fire, paper, or both.[33]

Now, given the ontology of mechanism, powers thus understood are hardly acceptable. But Malebranche holds on to two key features of the Aristotelian conception: a cause must logically necessitate its effect, and it must be directed at that effect by in some sense 'including' it. These features cannot be attributed to bodies as the mechanist understands them: on such a view, it would be a category mistake to think that an event, or a mode of a body, could include its effect. Similarly, as the 'no necessary connection' argument indicates, Malebranche thinks there is no logical impossibility in any conceivable sequence of sublunary events. Having ejected powers from the physical world, there is nothing in bodies that could ground such necessity. Thus, to satisfy these two requirements, Malebranche accords causal power only to God's will. A divine volition, as an act of an omnipotent being, is necessarily connected to its effects; equally important, a volition has a propositional content that is identical with that volition's effect. It

---

[33] It is important not to be distracted by the obvious fact that fire's burning paper depends on many other substances (e.g., oxygen). I have included only fire and paper in my statement of the position for the sake of brevity, but nothing turns on this.

is this content that differentiates volitions from bodies or their modes and allows them to have the connection envisioned by the Aristotelians. On the cognitive model, then, a true cause can logically necessitate its effect partly because it includes that effect as its propositional content. These facts, I shall argue, form the backbone of Malebranche's arguments for occasionalism.

Such a model is utterly at odds with one of the driving impulses of mechanism, namely, accounting for the behavior of bodies in terms of their mechanical qualities. In the context of the new ontology, the cognitive model leads inexorably to a top-down view. Thus, a quite different model of causation is developed in the works of Régis, Locke, and, to some degree, Boyle, which I call the 'geometrical' model. On this view, the necessary connection between cause and effect is analyzed in terms of the truths of geometry, which, for Locke, are synthetic and a priori. That a key can open a lock, for example, depends solely on the shapes of those two items. The tie that Malebranche thought could only be secured by intentionality is now secured by the intrinsic natures of the cause and the effect. It is an open question whether this new kind of connection is enough to fund logical necessity; indeed, this is precisely the question Hume will ask.

The divide between these cognitive and geometrical models of causation, which leads to the top-down and bottom-up models of law, will play a large role in defining the debate. Once we get the competing views in focus, we shall be in a much better position to understand Hume's contribution to the controversy, since only then will we be able to say with confidence precisely which positions he has in his sights.

2

Plan of the Book

Following these four themes through the admittedly narrow swath of the moderns I have carved out is a large task. I have chosen to break the project down into four main parts. Within each part, I proceed chronologically, and this is largely (though not always) true of the book as a whole.

Before we can begin to understand the modern debate, we must know something of the scholastic background against which it was conducted. Thus, in Chapter 3 I use two key figures of scholasticism—Aquinas and Suárez—to explicate the basic structure of Aristotelian thought on these issues.

In Part I, I sketch the Cartesian predicament. Having broken with the scholastics on key issues in ontology, Descartes faces a central tension. On one hand, he is clearly optimistic about the ability of mechanism to provide perspicuous explanations of body–body causation. On the other, he gives God such a primary role in the natural world that his activity seems to eclipse that of bodies. Descartes's God must re-create the world at every moment and, as a result, there seems little left for created beings to contribute. After all, if it is sufficient for motion that God re-create an object in a slightly different location at the next moment, it would seem that any activity on the part of objects is superfluous. This kind of observation forms the basis of one of Malebranche's central arguments for occasionalism. Another, and to my mind much more important, difficulty lies in God's role as the legislator of *lex naturalis*. A God who governs the actions of bodies in this way must, I shall argue, be directly and fully responsible for those actions. While we are accustomed to thinking of laws as governing events 'on their own,' such a view would have been all but unthinkable in a seventeenth-century context.

Much of the history of Cartesianism is the working out of these competing impulses, with the occasionalists playing up the role of God and figures like Régis struggling to find room for bodies as 'secondary causes' of motion in a way parallel to that found in scholastic concurrentism. I shall argue that, in his mature work, Descartes is committed to occasionalism in the case of body–body interaction. Although I of course think the considerations I advance are persuasive, I do not want to underestimate the ambiguity of Descartes's texts. Indeed, a figure like Descartes has the place he does in the history of thought partly because his work provides a number of different directions in which it can be refined. Thus, while I think the mature Descartes is an occasionalist with regard to body–body interactions, we should not ignore the echoes and strains of other views in his work.[1]

I then turn to an investigation of the concept of *vis* (force, power) in Descartes's work. It will emerge that Descartes regards force as a misleading way of pointing to one of the laws of motion, and that bodies thus do not have *vis*, strictly speaking. I then use these results, *inter alia*, to argue against a currently popular reading of Descartes that casts him as a concurrentist. In this part of the book, I canvass not only Descartes's invention of the laws of nature but his case against Aristotelian powers, since these arguments will set the bar against which later attempts to resuscitate powers are measured.

In Part II, I turn to the dialectic of occasionalism. I begin by exploring Malebranche's suite of arguments for occasionalism, some of which, as we shall see, have their origin in Descartes's own position. Malebranche's most intriguing arguments, however, take their cue from some features present in scholasticism. The 'being-toward' of powers, Descartes thinks, requires intentionality. Malebranche takes this to heart, and makes intentionality a necessary condition of causation. Malebranche presents what I have been calling the cognitive model of causation, which requires that cause and effect be linked by a kind of identity that he thinks only intentionality can provide. Not just any intentional state will do, of course; only volitions possess both intrinsic directedness and the potential for causal efficacy. The link between a cause and its effect is secured by the propositional content of a volition, since this is identical with its effect. This criterion serves not only to rule out bodies as causes but finite minds as well.

I then turn to the case of Pierre-Sylvain Régis, who fights a rearguard action by at once revising and preserving some central Aristotelian notions within the general context of Cartesianism. In particular, Régis defends concurrentism, with its attendant notions of form and power, and in the process modifies it to accord with the stark ontology of mechanism. This broad strategy—retaining and reinterpreting Aristotelian notions in a bid to resist the top-down influence of Descartes—will become a familiar one, as we encounter it again throughout Part III.

In Part III, I examine the Régis style of response to the Cartesian predicament as it appears in early modern empiricism. This bit of the story is further complicated by the host of specific (and sometimes purely empirical) problems Cudworth, Glanvill, More, and others raise for mechanism. By exploring these problems, and our philosophers' reactions to them, we can gauge precisely to

---

[1] Until recently, most readings of Descartes on body–body causation cast him as either an occasionalist (Garber) or a concurrentist (Clatterbaugh; Hattab). By contrast, Tad Schmaltz (2008) offers an ingenious reading of Descartes as a conservationist. While I agree with Schmaltz that Descartes's early views (in *Le Monde* and *Discours*) are best understood as a form of conservationism rather than concurrentism, and while Schmaltz and I agree that concurrentism is not Descartes's position, I shall argue that the *Principles* and later correspondence point at least to body–body occasionalism.

what degree and in what form these figures are committed to mechanism. For just as these problems are coming to the fore, the revival of powers in the context of the new ontology was beginning to bear fruit. Both Boyle and Locke attempt, to varying degrees, to preserve the notion of power and make it central to their respective conceptions of the structure and function of natural philosophy. Since Descartes's assault on embodied powers proved so influential in the modern period, it is illuminating to see how the much modified neo-Aristotelianism of Boyle and Locke fares against it. In this context, the details of the medieval debate over relations become important. Whether they know it or not, Locke and Boyle deploy arguments that were centuries old and are only intelligible in that context.

Finally, in this part I examine in some detail Locke's own version of mechanism, which has been the subject of much recent debate. I argue that, while not a concurrentist, Locke is, like Régis, an unrepentant Aristotelian when it comes to powers, causes, and laws of nature. Locke represents the clearest statement of the geometrical model of causation, a model that Hume takes as one of his chief targets.

In Part IV, I turn to Hume. Now, one can seemingly find two distinct Humes in the texts. The first is resolutely agnostic about the 'ultimate springs' and principles underlying phenomena; the second, with equal resolution, denies that there are or even could be such things as causation or substance, understood in the realists' sense. By exploring Hume's Newtonian project and its results, I show how to unite these two Humes into a single coherent figure. I then examine Hume's theory of meaning, since Hume takes language and mental representation to impose much narrower constraints on the range of intelligible positions than any of his empiricist predecessors. While hardly a 'positivist,' as many have tried to brand him, what I call Hume's 'semiotic empiricism' means that neither realism nor agnosticism with regard to causation is an option.

Hume's arguments against causal realism have been misunderstood partly because their proper targets have not so far been found. Once we have a firm grip on the cognitive and geometrical models of causation, both of which emerge as a consequence of the clash between (and melding of) mechanism and Aristotelianism, we are well on our way to seeing Hume's arrows home. Self-consciously responding to both models, Hume develops arguments that have their source in earlier writings but are transformed by their place in his novel method.

The clue to Hume's own position, I believe, is what I shall call 'the practicality requirement.' Causal reasoning is something we share with other animals and even children. The idea of a cause, and its most important constituent, the idea of necessity, are fruit that cannot fall but so far from the tree. (This partly accounts for Hume's otherwise puzzling failure to disentangle the questions of causation and induction.) Whatever our idea of causation is, it must have its source in a brute faculty of the mind, not in reasoning. Although such an idea, once had, can allow us to think of any two things or events under the concept of a cause, its source is ultimately in the workings of the imagination. Correspondingly, Hume develops two notions of relation, philosophical and natural, which mirror this twofold application. As a natural relation, causation includes a determination of the mind to move from one perception to another; as a philosophical relation, it omits this determination. I argue that Hume's two definitions of cause are not, and cannot be made, coextensive, since it is vital that he preserve a genuine sense of the term 'cause' that does not include this mental determination.

Hume, then, emerges as a figure who synthesizes a number of the themes I have devoted the rest of the book to exploring. Issues of intentionality, representation, the ontology of relations, and the nature of necessity will be woven together in our exploration of Hume's positive and negative views.

One key question will be with us throughout: why does the conception of causation as logical necessitation, understood as either the geometrical or cognitive models asserts, so long outlive its basis, the Aristotelian view of powers? I shall not be in a position to answer this until the very end of the book. But if I am right, the view lurks behind many of the otherwise unintelligible moves and positions in the debate. Indeed, it forms the basic view Hume attacks, almost a century after Descartes's *Meditations*.

3

The Aristotelian Background

The title of this chapter reflects a certain amount of insouiance. For there were not one but many Aristotelianisms on offer, both before and during the early modern period. Luckily, my purposes require only a brief sketch of some features of the most relevant and widely held positions. I shall draw principally on the *Metaphysical Disputations* of Francisco Suárez (1548–1617), as well as the much earlier work of Aquinas. The importance of Suárez for the study of not only late scholasticism but early modern philosophy generally is widely acknowledged; Suárez's *Disputations* are handy compendia of arguments and positions, and the views Suárez himself endorses were taken seriously by many outside his own intellectual community, the Jesuits. Attention to the curriculum prescribed to such figures as Locke and Descartes shows that both Aquinas and Suárez were among the authors that shaped their views.[1]

Late scholasticism is replete with so many dead ends and barely navigable avenues that it is tempting to think it was not gradually undermined by Galileo and Descartes et al. but collapsed under its own weight. Given my purposes in this book, the clearest way to begin is to work backwards, taking up the contemporary notion of nomological necessity and working to recover the Aristotelian view.

3.1 NECESSITY

As we have noted, a common contemporary response to Malebranche's (and Hume's) 'no necessary connection' argument is befuddlement: even if we accept that a given event is not logically necessary, why should anyone believe that it is not *nomologically* necessary, and hence not a bona fide instance of causation? Steven Nadler argues that Malebranche's identification of logical with causal necessity 'does seem strange today, and, I suggest, *should* have seemed strange to a seventeenth century Cartesian.' For between the eleventh and seventeenth century, 'there was a clear and dominant philosophical tendency

---

[1] This is confirmed by the curriculum common in 17th-century Cambridge and elsewhere. See Holdsworth (1648) and Trentman (1982). Often the influence was mediated by the presentation of Suárecian and Thomistic views in textbooks, such as that of Eustachius à Sancto Paolo. See Des Chene (1996: 11).

to distinguish causal or natural necessity—grounded in the operations of real efficient causes—from logical necessity.'[2] NNC is directed at a straw man.

But this should lead us to question, not Malebranche's or Hume's understanding of their philosophical adversaries, but our own. I shall argue that Malebranche and Hume in fact get it right: their immediate philosophical opponents, and most Aristotelians in particular, do indeed hold that whatever else causation may be, it requires logical, not nomological, necessitation.

The core position, traceable back to Aristotle, is based on the connection between a form and an object's powers. A substance does what it does in virtue of its form. That fire burns is an analytic truth, although one that can only be discovered through experience. Fire that failed to burn would, for that reason, simply not be fire. Ordinary transeunt actions, of course, require two substances, so to be precise, we should say that fire has the power to burn objects endowed with the appropriate passive power. We would no doubt have to circumscribe fire's capacity to burn with further statements of standard conditions, including the presence of oxygen, and so on. But at no point in those conditions would we have to enter anything like '. . . and assuming the laws of nature remain the same.' For the notion of a law in this contemporary sense is alien to the Aristotelian family of positions. Where the notion does appear, it is in the context of a divine command theory of ethics.[3]

This rough and ready characterization would need to be refined considerably to stand as an interpretation of Aristotle. But let us look instead to the scholastics. The chief difficulty they face, and one which will be with us for much of the book, is how to reconcile God's power and omnipotence with the powers of created beings. The dominant view, seen in both Aquinas and Suárez, is concurrentism.

Briefly, concurrentism holds that one and the same effect can be ascribed both to God and to natural agents.[4] God, as the primary cause, is responsible for the *esse* of individual beings; creatures, as the secondary cause, are responsible for the properties of those beings. Aquinas writes,

The order of effects is according to the order of causes. Now the first of all effects is being, for all others are determinations of being. Therefore being is the proper effect of the first agent, and all other agents produce it by the power of the first agent. Furthermore secondary agents which, as it were, particularize and determine the action of the first agent, produce, as their proper effects, the other perfections which determine being. (*SCG* 66: 119)

The typical metaphor by which Aquinas explains this curious dual contribution of God and secondary cause is that of a craftsman and a tool. The tool or instrument by itself does not produce, and is not a sufficient cause of, say, the wood being carved thus and so. Its power depends on the power of the craftsman

---

[2] Nadler (2000: 114). Nadler indicates in a footnote that he takes Aquinas to be among the exemplars of this tendency. See also Nadler (1996).   [3] But see below, Ch. 7.
[4] See *SCG* 70: 129–30.

using it. Nevertheless, that the wood is carved thus and so depends partly on the craftsman and partly on the instrument, for which instrument he uses, no less than how he moves his hands, will determine how the wood is shaped. 'The whole effect proceeds from [both God and the natural agent], yet in different ways, just as the whole of one and the same effect is ascribed to the instrument, and again the whole is ascribed to the principal agent' (*SCG* 70: 130). As we shall see, one of Aquinas's arguments against occasionalism is that if there were no true secondary causes, there would be no diversity in God's effects, since God is immutable. Secondary causes are required if God wishes to produce anything other than that which is, like him, immutable and uniform. That is, a primary and secondary cause are individually necessary and jointly sufficient for the production of any natural event, even though God retains his primacy as the source of all *esse*.

Unlike conservationism, which holds that God merely conserves bodies while their powers operate autonomously, concurrentism requires that God also, as it were, work through the powers of the objects he creates and conserves. And unlike occasionalism, which takes God to be the only real cause, concurrentism assigns genuine causal powers to objects. But how is it possible for an object to have this secondary power, or to serve as a secondary efficient cause, if God is nevertheless the ultimate source of all such power? Isn't this a case of overdetermination?

Suárez deals with this objection in the course of defending concurrentism from occasionalism. Suárez grants that overdetermination is impossible; that is, it is contradictory 'for the same action to proceed simultaneously from more than one total cause,' where 'total cause' refers to the sufficient condition for a given event. Unlike two total causes, however, the primary and secondary cause 'belong to different orders and are essentially ordered to one another' (*MD* 18. 1: 41). Just as an ordinary object exists in the fullest sense while depending on God for its existence, so an object's power can depend on that of God without being demoted to a power in name only.

It is a matter of dispute in the context of late scholasticism whether and in what sense a substantial form acts. Does it act by itself, or only through its accidents? A further question is whether there are accidents that are not mere instruments to the substantial form. Some of the issues here turn on precisely what one makes of the Eucharist.[5] The crucial claims for our purposes are these: whatever created being acts, acts only by virtue of God's concurrence; and created powers are either accidents alone (as in the case of the Eucharistic accidents), accidents that follow as a matter of necessity from the substantial form, or substantial forms themselves.

The natural world thus appears, much as it did to Aristotle, as a network of causal powers, the combination of which decides the outcome of any event. It is true that Aristotle often spoke as if the course of events had a bit of play in it, as

---

[5] See *MD* 18. 6: 128.

if it were not in what Simon Blackburn has called a 'straitjacket.' For Aristotle sometimes writes of that which happens 'always or for the most part,'[6] which suggests some flexibility in the nexus of powers.

This is easily reconciled with a thoroughgoing determinism, however. For when a given action fails to take place, the requisite passive power on the side of the patient might not have been present, as when one animal fails to impregnate another.[7] Indeed, the language of 'total cause' seems to have been developed precisely to capture these necessary conditions. While it does not logically entail determinism, the Aristotelian view as developed by Suárez and Aquinas is clearly consistent with it, at least in the realm of non-human phenomena.[8] (Again, I am leaving human and angelic agency out of my account.) For once the requisite active and passive powers are instantiated, and God concurs with these powers, it is a contradiction, and hence a logical impossibility, that the proper event not result. I should emphasize that, on my reading, the first relatum of the logically necessary connection is not the secondary cause alone but that cause plus God's concurrence. Concurrentists and occasionalists agree that without God's activity, no event can happen. Where they differ is on the cooperative causal role of creatures.

Let me explain the sense in which I am claiming that, on the present view, true causal propositions (claims such as fire burns paper, etc.) are analytic.[9] Although many notions of analyticity are available, the one that best captures the notion involved here is that according to which a proposition is analytic just in case its denial is a contradiction. One can still ask, in virtue of what does the contradiction arise? The answer is: the (fully adequate) concepts of the essences involved in the causal transaction. Here we move closer to a second notion of analyticity, that of conceptual containment. But for my purposes, the important claim is that the denial of a true causal proposition issues in a contradiction.

I must now defend this reading against two objections, one philosophical, one historical. The first objection is simply that in reading the scholastics as taking causal necessity to be but one species of logical necessity, I have done them a

6 See *Physics* 2. 5.
7 See Suárez, *MD* 19. 1: 281: 'natural causes can, as we have explained, impede one another through resistance or through a contrary action, and in this way they are also capable of removing all the things that are required for acting. But once these things have been posited, natural causes cannot prevent the action of a necessary agent, since they do not have the power to change the nature of things or to remove wholly intrinsic properties.' For more on the late scholastic debate, see Leijenhorst (2002: 182 ff.).
8 That every total cause necessarily brings about its effect does not entail that every event or object has a cause. Nevertheless, if we restrict ourselves to the sublunary and non-human realm, the latter claim is endorsed by the scholastics.
9 Note that the propositions we are considering are not spatiotemporally indexed; thus, absent further presuppositions, it is not analytically true that this match burned this paper *at 10:30 pm in Christiansburg Virginia*. Nevertheless, it is analytically true that fire burns paper; moreover, the conditional claim (if something with the essence of fire comes in contact with something with the essence of paper, burning will happen) is also analytically true. (These illustrations are abbreviated for the purposes of exposition; of course, the fully spelt-out proposition would be far more complex, mentioning not only God's activity but the passive powers of the relevant objects, as well. See below.)

disservice. For this turns their view into a bare tautology. If one packs everything needed to generate a given event into the putative cause, of course that cause will be generated; this is trivially true. By appealing to *ceteris paribus* clauses, cashed out in terms of the contiguous instantiation of the relevant passive powers, I have reduced what seemed like a bold causal hypothesis to an analytic truth. 'If everything necessary for fire to burn paper is present, then it will burn' is hardly informative. And philosophical objections like this one are often transformed into interpretive or historical objections: shouldn't we apply the principle of charity, and look for some other interpretation?

But the scholastic view as I have read it is anything tautological. Tautologies, whatever their faults, have the virtue of being necessary truths. But no Humean would regard it as a truth, still less a logically necessary one, that fire, plus or minus however many contributing conditions you like, will burn anything whatever. The sorts of claims I have in mind would be tautologies only if somewhere in the total cause they included a description like 'and anything else that is needed to bring about the effect.' But on the scholastic view, what else to include in our description of the total cause is a purely empirical matter. And so long as what gets included under the heading 'total cause' is not a blanket statement like the one above, it will not be tautological.

This is not to say that a true causal statement is not logically necessary. There is no possible world in which God concurs with a given power, the empowered object is in the presence of others with the requisite passive powers, and that power does not bring about its defining effect. For this is precisely what makes a power the power it is. The negation of a true causal claim is a contradiction.

There is another way to put this objection, using the second formulation of analyticity. Analytic propositions are knowable a priori because the (concept of) the predicate is contained in the (concept of) the subject. But it is hard to believe that the truths of natural science can be discovered by reflecting on our concepts. If causal claims were necessary in this way, natural science would be, as Stephen Mumford puts it, 'a trivially analytic human folly.'[10]

This point seems compelling only if we neglect the Aristotelian account of concept formation.[11] True causal claims are, on this view, a priori in the justificatory, not genetic, sense. Although the mind must undergo a complex set of experiences and operations to grasp the relevant concepts, causal claims are

---

[10] Mumford (1998: 237). Mumford's target is the view of Ellis and Lierse, who argue that the laws of nature are logically necessary because they are fixed by the dispositions of physical objects. Mumford claims that, like the Aristotelians, Ellis and Lierse take statements like 'x is an electron $\leftrightarrow x$ has behavior B' to be logically necessary. But, Mumford argues, this is deeply mistaken: 'That a particular possesses any disposition is logically contingent, even though some particulars, such as electrons, would not have been classed as such if they had different behavior. To deny this would be to claim that an electron's behavior is dictated by logic and, presumably, that physics is a trivially analytic human folly.'

[11] I am not, of course, suggesting that the scholastic view is right, or even a candidate for rightness.

ultimately justified by virtue of the connections between the essences involved, as captured in the abstracted concepts.

We get into a position to know causal claims not by stipulating definitions but by recognizing the true natures of the objects involved, a goal that can only be attained through repeated experience, under different conditions, of those objects.[12] Such experience allows the intellect to distinguish the complex of attributes essential to a thing's being what it is—its substantial form or organizing principle—from its accidental or nonessential characteristics.[13] The scientific concept of a natural kind is nothing but a more thorough and perspicuous working out of what was already present in the mind when it had initial perceptual contact with instances of that kind. Only the assumption that all analytic truths are true by stipulation or convention stands in the way of grasping these points.

Thus, despite being an empiricist, the Aristotelian does not think that concepts can be formed in anything like the *modern* empiricist fashion. When in perceptual contact with a thing, its intentional species is present in the mind. As Aquinas puts it, 'the intellect, according to its own mode, receives under conditions of immateriality and immobility the species of material and movable bodies' (*ST* I q. 84 a. 1). By considering the species of a given thing apart from its individuating conditions we can arrive at an understanding of its essence.[14] Nothing could be further from the post-Cartesian conception of experience and intentionality. This will have important consequences, particularly when we look at Hume's arguments against causal realism.

[12] To see this, we can briefly consider the scholastic conception of science, as sketched by Aquinas in his *Commentary on the Posterior Analytics*. *Scientia* is knowledge of 'that which cannot be otherwise' (*CPA* I. 9: 31). A demonstration makes a necessary conclusion known from necessary principles or premises, which themselves are better known than the conclusion (*CPA* I. 6: 24). Thus, the premises and conclusion of a demonstrative syllogism must contain predications per se ('i.e., in virtue of itself') (*CPA* I. 9: 32) and not per accidens, since the latter are not necessary (*CPA* I. 13: 43). Further, these predications must be universal, or 'said of all' (*CPA* I. 9: 32); if true, they will be true by de re definition (*CPA* I. 13: 43). To sum up: scientific knowledge consists in a body of syllogistic demonstration that moves from a set of necessarily true premises to an equally necessary conclusion.

[13] Aristotle writes, 'all animals . . . have a connate discriminatory capacity, which is called perception. And if perception is present in them, in some animals retention of the percept comes about, but in others it does not come about. Now for those in which it does not come about, there is no knowledge outside perceiving . . . but for some perceivers, it is possible to grasp it in their minds. And when many such things come about, then a difference comes about, so that some come to have an account from the retention of such things, and others do not. So from perception there comes memory, as we call it, and from memory . . . experience; for memories that are many in number form a single experience. And from experience, or from the whole universal that has come to rest in the soul . . . there comes a principle of skill and of understanding' (*Posterior Analytics* 99^{b}35–100^{a}9). See Aquinas, *CPA* II. 20: 237.

This point has been developed by Richard Sorabji in an effort to show that Aristotelian essences and the causal laws that flow from them are de re rather than analytic. See his (1980: 200 ff.). This is true enough if we take analyticity to amount to truth by stipulation, as I do not.

[14] *ST* I q. 85 a. 1.

At this point, a historical objection to my account of the scholastics might be raised. How, if true causal statements are logically necessary (because analytic), can God perform miracles? Surely most, if not all, scholastics are committed to the literal truth of the biblical miracles, as in the case of Daniel 3, where God prevents the fiery furnace from incinerating three young men, while the soldiers pursuing them burn?

The orthodox view is that although God cannot violate the law of non-contradiction, this hardly amounts to a denial of his omnipotence. But true causal claims are analytic, and so for God to bring it about that fire fails to burn would seem to involve just such a violation of the laws of logic. The challenge, then, is to reconcile God's ability to change the course of events with the 'straitjacket' view of nature. Concurrentism provides a handy way to do just this.

Suárez devotes *MD* 19. 1 to a discussion of 'causes that operate necessarily.' Unsurprisingly, this includes all beings except humans and perhaps angels. Now, once the requisite active and passive powers are in place, 'natural causes cannot prevent the action of a necessary agent, since they do not have the power to change the nature of things or to remove wholly intrinsic properties' (*MD* 19. 1: 281). Note what it would take here to prevent the action of such an agent, i.e., to change the course of events: one would have to alter its intrinsic properties. In other words, one would have to bring it about that fire was not fire.

Nor is there any exception for God. Although God *seems* to have this power, he really does not. God is able 'only to remove one of the required things.' God cannot bring it about that a natural cause fails to act when the conditions are right. When Shadrach, Meshach, and Abednego are lifted into Nebuchadnezzar's furnace, the soldiers who put them there are incinerated, while they themselves survive. But God did not remove the fire's power to burn, or flesh's passive power to be burned; all he did was withhold his ordinary concurrence from the fire.

For if God had decided on his own part to grant his concurrence and had left all the other required conditions intact, then he would have been unable to prevent the action. For *it involves a contradiction* to remove that which is natural in the absence of any contrary efficient causality, or at least without withholding the assistance or efficient causality that is required on God's part . . . And so once the presupposition in question, explained as above, has been made, the action arises with such a strong necessity that it cannot be impeded except by removing some part of what has been presupposed.   (*MD* 19. 1: 281–2; my emphasis)[15]

---

[15] One might object that the contradiction arises here simply because God cannot will both *p* and not-*p*, and thus that the created powers have no real role to play in explaining the impossibility of God's concurring with a power in the right conditions and yet the characteristic effect not taking place. I think this would be a mistake, however, since, as Suárez makes clear, the contradiction arises precisely in virtue of the nature of the created beings. It remains the case, however, that there is no logical contradiction in such a state of affairs if we subtract God's concurrence: fire that failed to burn flesh without God's activity would still be just the power it is. In addition, note that, for a

Just as Malebranche and Hume suppose, Suárez holds that it is logically contradictory, and hence inconceivable, that the presupposition of an action be present and yet that action fail to take place. God can remove his concurrence, but this is no different in kind from a situation in which the intended patient fails to possess the requisite passive power: part of the total cause is not present, and so the action cannot take place.[16] This remains the case, even though God is of course the pre-eminent factor in the total cause.

A central theme of the book is the persistence of this conception of the natural world as a nexus of logical necessities grounded in the natures, whether construed in Aristotelian or mechanist terms, of the bodies that possess them. Only by keeping this is mind can we see where the moderns diverge from the Aristotelians and where they fall in with them.

3.2 THE ONTOLOGY OF RELATIONS

The logical necessities that structure the physical world are grounded in powers, on the Aristotelian view. And whatever else they might be, powers are relations.[17] Their ontological status is thus one of the major fields of battle in the modern period, and one of the least explored.

There is no single position on relations that can be called 'orthodox' among the Aristotelians. The controversy over relations during the medieval period resulted in a bewildering variety of positions, including everything from a straightforward denial of their extra-mental existence to various kinds of reductive and non-reductive proposals. What I aim to set out here is a non-reductive realist position, one to which, as we shall see, Descartes, Boyle, and Locke, among others, are responding.

We can begin by noting a feature common to nearly all realist Aristotelian views.[18] While we are accustomed to treating propositions asserting relations as involving polyadic predicates (and hence polyadic properties, on a realist view), the Aristotelians are not. On their view, a perspicuous analysis of a relational statement would not take the form aRb. Theirs is a substance–accident ontology, and relations must somehow be fit into this. But how can a single property inhere in or unite two distinct subjects? As Ibn Sina argues, 'in no way may you think

concurrentist, God does not will states of affairs as such; instead, he achieves his effects *by* working through the powers of beings he has created. (I am grateful to a referee for the *Archiv für Geschichte der Philosophie* for pressing this and other objections.)

[16] 'Therefore, it is not the case that God brought it about that the fire did not act even though all the required things had been posited; instead, he removed one of those things' (*MD* 19. 1: 281).

[17] This is clearest in the case of transeunt actions, with which I shall mostly be concerned. But even the power to walk or to grow is a feature of an object related to and directed, if not at some numerically distinct object, then at some non-actual state of affairs.

[18] The following account is especially indebted to Henninger (1989).

that one accident is in two subjects.'[19] This sort of reasoning underlies various anti-realisms about relations, from Peter Aureoli and William of Ockham to Leibniz. For example, Leibniz argues that 'orders, or relations which join two monads, are not in one monad or the other, but equally well in both, that is, really in neither, but in the mind alone.'[20] Given a substance–accident ontology, if a relation aRb were an accident, it would have to be true that ($Ra. Rb$). But a token of an accident cannot be shared by two substances, for then 'we should have an accident in two subjects, with one leg in one and the other in the other, which is impossible.' Thus, a relation, 'being neither a subject nor an accident . . . must be a mere ideal thing.'[21]

The realist response to this is to split the relation down the middle into two distinct properties. Relations must in effect be pairs of accidents, each inhering in only one subject. Each relatum possesses a distinct property which itself somehow 'points to' or is directed at the other. So the proper form of a relational proposition is ($Ra. R'b$), where R and R' stand for what we shall call relational properties. These can be characterized by their *esse-in*, that is, their status as accidents in a subject, and their *esse-ad*, or 'being-toward.'

Many philosophers took this to be the natural reading of Aristotle's *Categories*, where Aristotle says that 'those things are relatives for which *being is the same as being somehow related to something*' (8^a33).[22] The being of a relational accident consists in this kind of directedness. So, to take an intrinsic relation as an example, *being lighter than brown* is a relational accident possessed by this yellow thing and is necessarily something other than its yellowness. The case is clearer with a monadic predicate like heaviness. Aquinas writes that 'in a heavy body is found an inclination and order to the center of the universe; and hence there exists in the heavy body a certain relation in regard to the center, and the same applies to other things' (*ST* I q. 28 a. 1). The inclination is not reducible to any further feature of the body, such as its size or shape. It is a real feature of the heavy body itself.

Not every relational property corresponds to a further such property in the object to which it is related. Some relations are merely verbal; thus, the Aristotelians typically distinguish between real relations, in which a relational property is involved, and merely verbal relations, in which it is not. Some relations are real only in one direction, as it were. The typical example here is intentionality. When I think about, say, the sun, my mind is really directed toward the sun in virtue of a relational property. But the sun is not really 'directed back' at my mind. To say that the sun is an object of my thought is not to report on any property of the sun at all. As John Sergeant puts it, 'the

---

[19] Quoted in Henninger (1989: 5).   [20] Letter to Des Bosses, May 29, 1716, AG 202.
[21] Letters to Clarke, AG 339.
[22] The interpretation of this text is not at all straightforward, and I would not want to be construed as offering an interpretation of Aristotle himself.

Objects [of sight or thought], for want of a *Real Ground*, are *not* really Related back to the Powers [of sight or understanding]; however the Words *Understood* or *Seen* do Verbally answer to the Acts of Understanding and Seeing; which is, therefore, call'd by the Schools in their barbarous Language *relatio des dici* or an Extrinsecal Denomination.'[23] Despite his derisive tone, and the fact that he is writing at the very end of the seventeenth century, Sergeant nevertheless articulates a thoroughly orthodox scholastic view.

The natural world, then, on Sergeant's Aristotelian view, is a network structured by powers and relations, each of which has not just some real foundation or justification in reality but a positive ontological status of its own. Powers are of course the most important relations for understanding the natural world, since knowledge of them would allow us to predict precisely what is about to happen.

Although the view that powers are relational properties in this sense sounds odd, its recognizable descendant has surfaced in the contemporary debate over scientific essentialism. Brian Ellis offers a view according to which the properties of particulars are one and all powers or dispositions to bring about certain effects. Properties, then, are 'threats and promises';[24] they have their implications for future and possible states of affairs written into them, since this is precisely how they are to be individuated, both metaphysically and epistemically. (Sydney Shoemaker suggests a broadly similar position, as do Harré and Madden.[25]) But what exactly is going on when the power is not reduced to act? As D. M. Armstrong puts it in his response to Ellis:

in general, at least, the particular will not *manifest* this power at all times. Because its power constitutes the essence of the property on Ellis's theory, then if it does manifest the power, then it manifests that power *of necessity*... Suppose the manifestation does not occur. The causal power that the particular still has remains essentially one that 'points' to that manifestation, even where the manifestation does not occur. George Molnar speaks of this as 'physical intentionality,' and accepts that it is real. Ellis, I maintain, has to accept this also, as should Shoemaker, Swoyer, and Martin.[26]

While not thinking that a refutation is in the offing, Armstrong finds this unacceptable. 'A relation to a non-existent object!'[27]—even Meinong might hesitate. In reply, Ellis argues that his powers are not directed at specific non-actual events but at *kinds* of events or processes.[28] On its face, this reply does

---

[23] Sergeant (1696: 254).   [24] As Stephen Mumford (1998) puts it.

[25] See Shoemaker (1980) and Harré and Madden (1975).

[26] Armstrong (1999: 35). It is worth just noting that in an earlier paper, Armstrong invokes physical intentionality to explain the intentionality of mental states. Adducing the example of poison, Armstrong asks, 'Does it not provide us with a miniature and unsophisticated model for the intentionality of mental states? Poisons are substances apt to make organisms sicken and die when the poison is administered. So it may be said that this is what poisons "point" to' (Armstrong 2002).   [27] Armstrong (2002).

[28] Ellis (1999*b*: 40). A more promising line of defense might be George Molnar's (2003: 62 ff.) suggestion that intentionality is not a real or 'genuine' relation, since these require the existence

not help much. If it is mysterious how a power can point to a single, non-actual event, it is no less mysterious how it could point to a kind or class of those events. If kinds are reduced to sets of individuals, then this only exacerbates the problem, even if some members of the set exist in the actual world; if kinds are not so reduced, we have a case of a particular power being intrinsically directed at universals, or sets of tropes.

I do not claim that Armstrong is right here. I do think, though, that his sort of worry is what animates the early modern critics of Aristotelianism I shall discuss. Nor is George Molnar's talk of 'physical intentionality' without its historical counterpart.[29] For one of the chief arguments deployed by Descartes and Malebranche against their Aristotelian opponents was that the *esse-ad* of powers is a feature only minds possess.

3.3 MANIFEST AND OCCULT QUALITIES

A final barrier to a clear grasp of the debate concerns occult qualities. The moderns, on the surface, seem to speak with a single voice in condemning such qualities. But the issue is not so simple.[30]

In the context of scholastic Aristotelianism, occult qualities are defined in contrast to manifest qualities. The primacy of sensation restricts explanatory value to those qualities that are present to the senses. Since the intellect works by abstracting the intelligible species from the phantasm or image, scientific investigation is limited to those qualities that can be given in sensation.[31] A fire's heat, for example, is a manifest quality, which explains and causes the heating and burning of those objects with which it comes into contact. Such a quality is also a terminus of explanation: once the manifest quality has been found, inquiry can come to an end. An occult quality, then, is one that is (*a*) insensible, and for that very reason (*b*) unintelligible; it is thus (*c*) a terminus of investigation since, *ex hypothesi*, the intellect can go no further in exploring it. Paradigm cases of occult qualities include magnetism and the workings of the planets in human life.

of their relata. In this Molnar follows Brentano's own line; but if one is worried about ontological slums at all, those populated by pseudo- or quasi-relations should seem just as troubling as the non-existent relata themselves.

[29] Molnar argues in detail that intentionality can be possessed by the non-mental. In particular, he thinks that the directedness of physical intentionality need not be mediated by any sort of representation. See esp. Molnar (2003, §3).

[30] As Keith Hutchison (1982) shows; his article is the most useful I have found on the occult–manifest distinction, and the moderns' reaction to it. See also Clarke (1989).

[31] For example, Daniel Sennert writes in 1632 that 'occult or hidden Qualities are those, which are not immediately known to the Sences, but their force is perceived mediately by the Effect, but their power of acting is unknown. So we see the Load-Stone draw the Iron, but the power of drawing is to us hidden and not perceived by the Sences' (*Thirteen Books of Natural Philosophy*, quoted in Hutchison 1982: 234).

When the moderns object to occult qualities, what are they really worried about? Not (*a*), surely, since many versions of mechanism posit insensible causes for observable effects.[32] (And any moderns who turn out to be indirect realists will *a fortiori* deny the direct observability of *any* qualities.) The real issue here centers on features (*b*) and (*c*), and their link to (*a*). The scholastics' epistemology, founding all *scientia* as it does on manifest qualities, rules out appeal to insensible qualities as sources of knowledge; worse, it provides an excuse for giving up, since, being insensible, these qualities are therefore unintelligible, and inquiry can go no further. Thus, Walter Charleton in 1654 inveighs against 'Secret Sympathies and Antipathies: for as much as those Windy Terms are no less a Refuge for the Idle and Ignorant, than that of Occult Properties.'[33]

When Joseph Glanvill comes to attack the doctrine of occult qualities in his 1665 work *Scepsis Scientifica*, he clearly has (*c*) in mind. As to nature's 'more *mysterious* reserves, *Peripatetick* enquiry hath left them unattempted; and the most forward notional Dictators sit down here in a contented ignorance: and as if nothing more were knowable then is already discover'd, they put stop to all endeavours of their Solution.'[34] And when he argues that the Aristotelians 'resolve all things into *occult* qualities,' he means, not that they appeal to insensible qualities in their explanations, as almost any mechanist would, but that they fail to provide any explanation whatsoever, even of the qualities they themselves deem manifest. Thus, early modern objections to occult qualities must be seen as the new philosophy straining against the epistemic limits of the old and insisting on the in principle intelligibility of such qualities as magnetism, which had previously been thought beyond the reach of human knowledge.[35] If the world is 'an America of secrets,'[36] it is nevertheless ripe for exploration.

It would be a mistake to see occultism as a charge leveled solely against the Aristotelians. Throughout the modern period, occultism is a bugbear. Most famously, of course, Leibniz charges Newton with making gravity an occult power. More interesting for our purposes, however, is the fact that Newton and his champion Roger Cotes level this charge against many of the mechanists themselves. As we shall see, some mechanists, Descartes, Malebranche, Régis, and perhaps even Locke among them, 'take the liberty of imagining that the unknown shapes and sizes of particles are whatever they please . . . and even further of feigning certain occult fluids that permeate the pores of bodies very freely.' Even if these philosophers 'proceed most rigorously according to

---

[32] It is tempting to think that an occult quality for the moderns generally means a quality that is not just insensible but one that also has no mechanical analog. While I do not deny that this use occurs in the modern period, it is too limiting to serve as a general model of the concept, since Newton and Cotes, for example, object to the unbridled positing of insensible mechanical qualities. See below.　　　　　[33] Charleton (1654: 344), quoted in Hutchison (1982: 244).

[34] Glanvill (1665: 125).

[35] For a nice example of this, see Descartes's *Principles*, iv. 187 (AT ix. 309).

[36] As Glanvill liked to put it. (Though Glanvill would not agree with the sentiment expressed at the end of this sentence.)

mechanical laws,' they are 'merely putting together a romance, elegant perhaps and charming, but nevertheless a romance.'[37]

We now have in place a view that deserves to be called the dominant late scholastic position. Its four key features are these:

1. It is a bottom-up view. That is, its fundamental notion is power, not law. Events happen as they do because of the natures of created beings and God's concurrence.

2. A true causal statement is analytic; its denial issues in a contradiction. This is so even though the total cause of an event now involves God's concurrence, as it did not for Aristotle himself.

3. At least on some prominent Aristotelian views, powers (and relations generally) are accidents, not polyadic predicates or universals, and are not reducible to their foundations in beings.

4. Some powers of objects, namely, their manifest qualities, are capable of being made present to the senses and hence in principle intelligible; others are not.

This outline will be filled in as we go. For Aristotelianism does not merely form the background to the modern debate; in various guises, it is part of that debate itself. Many of the key positions that serve to differentiate the Aristotelianisms on offer in the late scholastic period recur in the modern. Indeed, on some issues, nominally anti-Aristotelians adopt positions that are recognizable within earlier scholastic debates, the parties to which all considered themselves faithful interpreters of the Philosopher. Thus, Sergeant defends a non-reductive realism about relations and hence powers, previously defended by Albertus Magnus and Aquinas, which Locke attacks, using arguments that were first put forward by William of Ockham and Peter Aureoli. I thus hope to sketch a considerably more nuanced and (perhaps regrettably) complex picture of the role of broadly Aristotelian thought in the modern debate than is usually offered.

---

[37] Cotes (2004: 43). See Clarke (1989), who titles one of his chapters on the Cartesians 'Hypotheses Fingo.' Newton himself parries the charge of occultism by appealing to his distinction between mathematical and physical hypotheses; he is concerned only with the former.

PART I

THE CARTESIAN PREDICAMENT

4

What Mechanism Isn't

What makes an early modern philosopher a 'mechanist'? There is no doubt that mechanism begins with the idea that the natural world can be understood on analogy with a machine. There should be just as much prospect of understanding the workings of the kidneys, for example, as there is of understanding the workings of the famous clock at Strasbourg, or the water-driven statues Descartes was so taken with at St Germain-en-Laye.[1] But how precisely is this analogy to be understood?

Consider Michael Ayers's characterization of mechanism as the view that 'the laws of physics can be explained, in principle if not by us, by being deduced from the attributes possessed essentially by all bodies *qua* bodies, i.e., from the nature or essence of the uniform substance, matter, of which all bodies are composed.'[2] Mechanism, for Ayers, is primarily a position on the laws of nature; what is more, it is a bottom-up conception of these laws, since they are supposed to be in principle deducible from the properties of matter.

If we test this definition against two benchmarks of seventeenth-century mechanism, namely, René Descartes and John Locke, we find that it fits neither. On Descartes's view, as we shall see, the laws of physics derive primarily from God's immutability, with the nature of matter playing a distinctly limited role, if any. A natural philosopher might know all there is to know about the essential attributes of bodies and yet be as far from knowledge of the laws that characterize their behavior as a toddler.

In a quite different way, Ayers's characterization misses Locke. For Locke, like his friend Robert Boyle, eschews talk of 'laws of nature' (except, of course, where that phrase is meant in its moral sense, as in divine command theory). I can only find one place in all of the *Essay* (IV. iii. 29: 560) where Locke uses 'law' in anything like its extra-moral sense. Nor is this merely a terminological matter: what counts for Locke is not law but power. If we have to talk in terms of laws at all, they will be statements of regularities that depend for their truth on the real essences of corporeal substances.

---

[1] See Descartes's *Traité de l'Homme*.

[2] Ayers (1981: 210). In later works, Ayers went on to modify his definition of mechanism. I have chosen this earlier formulation for its value as a foil in developing my own.

Other definitions have been proposed. At the risk of introducing some neologisms, I think clarity is best served by starting over. It seems to me we might mean two quite different things by 'mechanism.' First, we might mean the view that the only properties of bodies are size, shape, and motion or rest, plus or minus derivative properties like situation, position, and so on.[3] This is what I shall call 'ontological mechanism.'[4] Many, but far from all, of the figures I shall be concerned with are mechanists in this sense. It is important to note that, while ontological mechanism might have consequences for the issues of powers and laws of nature, it is not in the first instance a position *on* these issues.

To capture these other elements, I shall speak of 'course of nature mechanism.' This is the view that physical objects behave as they do in virtue of their properties and natures. Thus, an ideal observer like Laplace's demon who knew all of the relevant properties would be in a position to deduce precisely what was about to happen. Course of nature mechanism is a bottom-up picture; what happens, happens because objects are what and how they are, not because we happen to occupy a world where the laws of nature are thus and so, or because God directly wills it. More precisely, we can say that course of nature mechanism is the view that, given a state of affairs that specifies all the properties of bodies, including the direction and quantity of motion, the immediately following state of affairs comes about of necessity.

While both views deserve the title 'mechanism,' they are an unhappy match. As we shall see, they are connected or severed by distinct tissues of argument in each philosopher we shall examine. The modern debate begins as a reaction to the alleged ontological excesses of the scholastics. This reaction results in the wholesale rejection or, better, transformation of what had been an absolutely central notion in natural philosophy: power. And in the work of Descartes and some of his fellow travelers, the reaction results in a strong preference for top-down views. For these philosophers, to adopt ontological mechanism was to reject course of nature mechanism. So far from entailing the position that mechanical properties are fundamental to the course of nature and alive with causal efficacy, ontological mechanism in its earliest form encourages precisely the opposite attitude.

---

[3] The list of these properties varies considerably from one author to another, and even within the same author. Even Locke, in the space of one chapter, includes different elements in his list of primary qualities; see *Essay*, II. viii. 8: 135 and II. viii. 26: 142–3.

[4] It is tempting to add a justification for this selection of properties and claim that members of the privileged set are *mechanical* in the sense that they are relevant to the interactions of bodies with one another. Ontological mechanism would thus by definition rule out secondary qualities, since things like color and taste will not, of themselves, be relevant to the explanation of the motion of bodies, although of course the primary qualities that, on some views, cause us to experience the relevant qualia will. Unfortunately, this fuller definition of ontological mechanism leaves out such obvious candidates as Malebranche, since he, of course, denies that bodies genuinely interact, if we mean this to imply that they have at least some causal efficacy. Thus, I shall keep to the narrower of the two meanings.

There is no little irony in this, since ontological mechanism's initial impetus was precisely to provide a more perspicuous model of causal explanation. But by rejecting powers, many of the earliest defenders of ontological mechanism were in danger of banishing causation out of the natural world.[5] And in the case of Malebranche, a thoroughgoing ontological mechanism coupled with a traditional insistence on the ubiquity of God's causal powers results in the severing of causation from explanation altogether.

In what follows, I begin by examining some of the arguments, explicit and implicit, deployed by Descartes and others to undermine the Aristotelian bottom-up picture. These arguments set the stage for the attempts we shall examine later in the book (in Parts II and III) to resuscitate the scholastic notion of power. It is primarily in response to them that such otherwise quite different philosophers as Régis, Boyle, and Locke undertake their various reductions of power. What is of more immediate importance, however, is the light they might shed on Descartes's conception of God's causal powers. This is the burden of Chapter 5.

In Chapter 6, I turn to Descartes's ontology. His substance–mode ontology will provide an important constraint on the interpretation of *vis* (force or power) in Descartes's physics. But I shall also argue that Descartes's conception of the relation between substance, essence, and mode has consequences for any reading of his position on occasionalism.

The laws of nature and their derivation are the focus of Chapter 7. While not altogether novel, Descartes's picture of the nature of laws generally, and of motion in particular, represents the key turning point away from the power-driven physics. For Descartes explicitly replaces the scholastics' secondary causes with the laws of nature. While initially quite puzzling, I show how Cartesian laws can serve as functional replacements for scholastic secondary causes. The following chapter provides a reductive analysis of *vis* in Descartes's mature physics.

In Chapter 9, I trace out the consequences of these positions for the question of causation. There are four options on the table. First is conservationism, a rather unpopular[6] view that takes God's activity in the created world to consist merely in keeping beings in existence; the changes that happen in those beings, by contrast, are due to their own causal activity or that of other beings. Next we have concurrentism, according to which, as we have seen, God works through creatures to produce effects. At the third point on the scale we have limited occasionalism, which argues that, while God is the only true cause of body–body

---

[5] Ayers (1996) seems to have something like this tension in mind.

[6] Freddoso points out that 'almost all of the important figures in the history of philosophical theology have rejected [mere conservationism] as theologically "unsafe"' (1991: 24) and quotes Albert the Great, who wrote that conservationism 'has all but disappeared from the lecture hall.'

interaction, minds have causal powers. Finally, of course, there is full-blown occasionalism, denying causal powers to all created beings.

I shall argue that Descartes's view undergoes a transformation. I argue that in early works such as *Le Monde*, Descartes finds a version of the conservationist option appealing. Despite its superficial similarities to concurrentism, this early view is really a conservationist one; the differences will become most stark when we contrast this early view of Descartes's with a genuine mechanistic concurrentism, that of Pierre-Sylvain Régis. As we shall see, Descartes shows signs of increasing awareness of the tensions between his view and conservationism (and indeed concurrentism); Descartes, on my view, changes his mind in his mature works, such as the *Principles* and *Meditations*, which are self-consciously committed to body–body occasionalism.[7] To prove this, I develop an argument from the laws of nature, prefigured in Cudworth, to show that concurrentism, arguably the most popular contemporary reading, is simply incompatible with Descartes's picture of the origin and force of laws. It then becomes clear that the mature Descartes is a limited occasionalist, in the sense that he at least denies causal powers to bodies. Thus, my central, though not sole, argument, can be summed up in the following way: conservationism and concurrentism are bottom-up views, while the mature Descartes cleaves to a staunchly top-down picture. But at this point we run up against what has become known as the 'divine concursus argument,' formulated by later Cartesians and sometimes attributed to Descartes himself, which threatens to deny all causal power to any beings other than God. Although Descartes clearly does not wish to endorse a thoroughgoing occasionalism, I argue that his view lacks the resources to answer this argument. In particular, some of the ways out favored by other commentators—distinguishing between the preservation of a substance and that of its modes, for example—run afoul of Descartes's ontology. I conclude Part I by examining the problem of mind–body causation and its implications for our conception of how God acts in the world. Descartes emerges as a reluctant occasionalist, committed to the denial of causality in the sublunary world altogether.

---

[7] Indirect evidence for my developmental view comes from Garber (1993: 19–24), who argues that Descartes's position on body–mind causation undergoes a similar transformation. In the *Meditations*, Descartes seems committed to the claim that bodies act on minds, whereas the French version of the *Principles* moves to a limited occasionalism, with bodies only serving as the occasions for God to produce sensations in us. I shall not repeat Garber's arguments here, since I think he has it right.

5

The Rejection of Aristotelianism

What defines ontological mechanism as I have characterized it is largely what it rules *out*: Aristotelian forms and powers. Let us begin, then, by looking at these notions as they were understood by Descartes. One caveat: I shall not consider the degree to which Descartes got the Aristotelian view right. Happily, Descartes paints with so broad a brush that at least quite a number of the disparate views flying under the banner of 'Aristotelianism' will be covered.[1]

It is not always clear precisely which Aristotelian notion Descartes has in his sights. He tends to lump their notions of quality, form, and power together, usually in the course of condemning the lot of them. Thus, he invites Morin to compare his 'single assumption that all bodies are composed of parts' with 'all [the scholastics'] *real qualities*, their *substantial forms*, their *elements*, and countless other such things' (July 13, 1638, AT ii. 199/CSMK 107). Nevertheless, we can make some distinctions here. Descartes takes his opponents to think that at least some of the qualities of bodies are 'real qualities,' qualities that 'have an existence distinct from that of bodies' (Letter to Elizabeth, May 21, 1643, AT iii. 667/CSMK 219). Descartes also speaks of 'occult qualities,' qualities that are neither sensible nor intelligible. Such occult qualities are unobservable powers that explain the observable behaviors of the objects in which they inhere. Although real and occult qualities need not be coextensive, since the former embodies a metaphysical claim, and the latter, an epistemic one, Descartes seems to run them together. Moreover, Descartes seems to use form and quality interchangeably, and, as in the passage below, treats these either as powers or as that which determines the powers of an object.

Descartes's most common argument is that qualities or forms are explanatorily impotent. This is explicit in *Le Monde*, where Descartes discusses the Aristotelian understanding of the process of combustion:

Others may, if they wish, imagine the form of fire, the quality of heat, and the process of burning to be completely different things in the wood. For my part, I am afraid of mistakenly supposing there is anything more in the wood than what I see must necessarily be in it, and so I am content to limit my conception to the motion of parts. For you may posit 'fire' and 'heat' in the wood, and make it burn as much as you please: but if

---

[1] A number of divergent Aristotelian views will be discussed in Parts II and III below.

you do not suppose in addition that some of its parts move about and detach themselves from their neighbours, I cannot imagine it undergoing any alteration or change.   (AT xi. 7/CSM i. 83)

What Daniel Garber calls the 'argument from obscurity' is implicit here as well. Often, Descartes simply claims not to understand what the scholastics mean when they speak of forms and qualities (cf. Letter to Henricus Regius, January 1642, AT iii. 506/CSMK 208–9). This kind of objection is of course clearest when directed at the doctrine of occult qualities (cf. Letter to Mersenne, AT iii. 649/CSMK 216, where Descartes calls our inability to conceive of these qualities his 'principal reason' for rejecting them). On the other hand, if this is to be more than mere calumny, Descartes must give some positive reason for regarding the scholastics' notions as irredeemably obscure. Now, Descartes offers a number of arguments to show that the essence of bodies is nothing but extension. Matter can be clearly and distinctly conceived with the resources provided by the idea of extension alone; anything else is otiose. This argument from elimination might be implicit in the passage above, since the essence of a body Descartes goes on to explicate is precisely 'what [he] see[s] must necessarily be in it,' and this, of course, is simply extension.[2] He refrains from attributing forms and qualities to objects because these concepts will have no place in a clear and distinct idea of body. This is even more straightforward in the case of real qualities, since they are supposed to be capable of independent existence and thus fit the traditional definition of substance while also inhering, most of the time anyway, in a substance, and hence fitting the definition of a quality.

For the sake of convenience, we can generate a list of the unattractive features powers allegedly have, since this will make it easier later on to see how other philosophers develop their own notions of power. Thus far, we have seen that qualities and their powers are, as Descartes reads the scholastics, character-ized by:

(1) *Ontological independence.* Aristotelian qualities are not reducible to, and do not supervene on, the mechanical properties (size, shape, and movement) of the objects they belong to. This is especially clear if we are thinking of real qualities, which, by definition, can come loose from their moorings, the substances they inhere in. But in any event, qualities are supposed to be more than matter, whether understood in the Aristotelian or Cartesian sense. This, of course, is the feature targeted by the argument from obscurity. When we turn to Descartes's own view, we shall see that a substance's modes just are that substance itself existing in a certain way; as Malebranche was to put it, modes are a thing's *manière d'être*.

---

[2] I should note that the wax argument of the *Meditations* does not rule out flexibility and mutability as candidates for the essence of the wax; but the implication, given Descartes's overall view, is that these dispositions follow from its extended nature.

(2) *Explanatory impotence.*[3] Since even the scholastics admit that those qualities chiefly invoked to explain observable changes, as something over and above the mechanical properties of matter, 'are occult, and that they do not understand them themselves,' 'these forms are not to be introduced to explain the causes of natural actions' (Letter to Regius, January 1642, AT iii. 506/CSMK 208–9).

These arguments from obscurity and explanatory impotence are of course central to Descartes's project and would be repeated throughout the modern era in polemics against the Aristotelians. What is more important for our purposes, however, is his third argument, which Dennis Des Chene has dubbed 'the little souls argument.'

Descartes writes that he does 'not suppose there are in nature any *real qualities*, which are attached to substances, like so many little souls to their bodies, and which are separable from them by divine power' (Letter to Mersenne, AT iii. 648/CSMK 216). In his *Sixth Replies* and elsewhere (e.g., Letter to Arnauld, AT v. 222–3/CSMK 358), Descartes accuses the scholastics of anthropomorphizing nature, in so far as they use concepts derived from the activity of the human mind in explicating natural phenomena. In a perhaps disingenuous bit of intellectual autobiography, Descartes confesses that at one time he similarly conceived of heaviness as a real quality inhering in bodies, over and above their mechanical qualities, just as Aquinas does.[4] 'And although I imagined heaviness to be diffused throughout the heavy body, I did not attribute to it the same extension which constitutes the nature of body.' What is more, 'while [heaviness] remained coextensive with the heavy body, I saw that it could exercise its force in any part whatsoever' (AT vii. 441–2/CSM ii. 297). A quality like heaviness is thus like a Cartesian soul in that it is diffused throughout the entire body that possesses it. But the final and most important objection is that the scholastics have projected the intentional capacities of the mind onto the physical world. Descartes writes, 'I thought that heaviness bore bodies toward the center of the Earth as if it contained in itself some knowledge of it.' This is what 'most especially shows that I derived my idea of heaviness from the idea of my mind.'

This last feature of the little souls argument was picked up and exploited by Malebranche. In arguing that bodies cannot have the power to move themselves, Malebranche writes,

Well, then, let us suppose that this chair can move itself: which way will it go? With what velocity? At what time will it take it into its head to move? You would have to give the chair an intellect and a will capable of determining itself. You would have, in short, to make a man out of your armchair.[5]

---

[3] Garber calls this the argument from parsimony. This is fine; I prefer my tag only because it brings out the fact that for Descartes, these notions are ontologically otiose *because* they are explanatorily otiose.   [4] See above, sect. 3.2.

[5] *Dialogue* VII, in Malebranche (1992: 227).

To sum up the little souls argument, we can say that Aristotelian powers are alleged to have the following unattractive features:

(3) *Physical intentionality.* Like a mind, a body endowed with power would have to be intrinsically directed at states of affairs. Moreover, these states of affairs need never be actual: fire would have the power to burn paper even if it never actually did. Although inhering in a single object, the power is directed toward a range of non-actual states of affairs. This *esse-ad* is the target of Descartes's claim that a body endowed with heaviness, conceived as a power or quality, would have to know where the center of the earth was if indeed it genuinely tended, of its own volition as it were, toward that location. And Malebranche rejects this feature by saying that power attributions require attributions of intellect. In both cases, it is the property of intentionality that is crucial, and this is a property both figures agree can be possessed only by minds.

This line of thought has often seemed like a deliberate misrepresentation of the scholastic view. While they often characterized powers as involving intentionality, they clearly did not think that bodies were possessed of little souls.[6] Nevertheless, one dominant scholastic view does indeed hold that an object's intrinsic properties, namely, its powers, point to or are directed at other objects and states of affairs, just as representational mental states can be. In addition to the above considerations, we can add that Aristotelian powers are after all supposed to link the agent and patient in a logically necessary way. So far from being 'entirely loose and separate,' as Hume will later put it, two objects or events, when connected causally, are also connected logically. As we shall see below (Chapter 10), Malebranche in effect argues that physical events are not of the right ontological category to serve as causal relata; this is because he retains enough of the scholastic picture to be committed to causation as logical necessitation, but not enough of their ontology to be able to conceive of anything but volitions and their effects as causes.

(4) *Physical teleology.* Empowered bodies would of themselves take certain states or events as their goals. They would be 'masters of themselves,' having, like human bodies, souls attached to them capable of directing their endeavors. This is of course connected with the previous point, since the doctrine of final causes requires that a given object, of its own nature, point to or be directed at its natural future state. The claim that each thing has a *telos* goes further than this, however. Again, both Descartes

---

[6] Nor did they all agree that a being's *esse-ad* represented a mind-independent feature of the world. We shall explore these views in greater detail below (Ch. 16).

and Malebranche believe such a relation can only be understood in mentalistic terms.[7]

These four features were to define the philosophical landscape for much of the next century. Philosophers such as Régis and Locke who wish to resurrect powers must either denude them of these features or find a way to make them acceptable in the context of the new, stripped-down ontology of mechanism.

What is more important for our immediate purposes, there might be a bit of baby in Descartes's bathwater. That is, in rejecting the Aristotelian view, Descartes, as we shall see, finds himself driven to a very strong top-down view that makes explaining the antics of ordinary physical objects in terms of their qualities all but impossible. But before we can investigate his position on the laws of nature, we need to have Descartes's own ontology before us.

---

[7] Cf. Des Chene (1996: 394). On the intrinsic directedness of Aristotelian powers, see esp. Mark Henninger (1989: 5 ff.), and Julius Weinberg (1965); for a contemporary endorsement of physical intentionality, see Molnar (2003).

6

The Nude Wax: Cartesian Ontology

Descartes's conception of the material world was taken up by Malebranche, Régis, and many others. And where philosophers of the time disagree with Descartes, they typically challenge his denial of the void or the thesis that matter is infinitely divisible, issues that are not directly relevant to our main concern. Descartes's metaphysics, particularly his conception of the relation between substance and mode, will constrain any interpretation of his position on causation. I shall argue below, for example, that a prominent interpretation that takes Descartes to be a concurrentist founders precisely on these shoals. Equally important are the problems that Descartes's ontology generates for the concept of *vis* (force, or power), a concept Descartes helps himself to in the *Principles*, even though, as we shall see, there seems to be no way to make room for it in his substance–mode analysis of material substance.

To begin, we must note that Descartes is a nominalist in the relatively uninformative sense that, for him, everything existing outside of the mind is particular.[1] And each particular is either an individual substance or a mode of a substance (AT viiiA. 23/CSM i. 208; cf. AT ixB. 45). A mode of a substance must also be individual: he speaks of modes as 'concrete' rather than 'abstract' (AT iii. 356/CSMK 178).[2] Universals, by contrast, exist only in the mind; they arise 'solely from the fact that we make use of one and the same idea for thinking of all individual items which resemble each other' (AT viiiA. 27/CSM i. 212).

Our grasp of the essence of any substance, however, does not consist in our generating such a universal. Instead, Descartes claims that such understanding comes about when we apprehend the 'principal property that constitutes [the substance's] nature and essence, and to which all its properties are referred' (AT viiiA. 25/CSM i. 210). The essence of material substance is extension, 'the property . . . of taking up space' (AT xi. 35/CSM i. 92), while that of mental substance is thought.

---

[1] I call this uninformative because it casts too wide a net. Aristotle, at least in some texts, would meet this definition of nominalism; so might Plato, if we take seriously his suggestion in the *Timaeus* that the forms are exemplars and hence particulars.

[2] Although Descartes typically reserves 'attribute' for the essence of a substance, he does say, for example, that 'by *mode* . . . we understand exactly the same as what is elsewhere meant by *attribute* or *quality*' (*Principles*, i. 56, AT viiiA. 26/CSM i. 211). For my part, I always use 'property' in a neutral sense to refer to any feature, whether mode, attribute, quality, *propria*, or power.

A good way into Descartes's metaphysics of the natural world is the famous piece of wax argument in Meditation II. As I read the argument, its conclusion is that no determinate properties of the wax belong to its essence, and so that essence can only be apprehended by the intellect.[3] Descartes suggests that we begin by 'tak[ing] away everything which does not belong to the wax and see what is left' (AT viii. 31/CSM ii. 20). Observed at time t, the wax has determinate feature F_1; at t', it has F_2, and could take on $F_3 \ldots F_n$. Descartes then asks whether sensation or imagination could be responsible for his clear and distinct idea of the wax. The answer, of course, is no, because 'I would not be making a correct judgment about the nature of the wax unless I believed it capable of being extended in many more ways than I will ever encompass in my imagination' (AT viii. 31/CSM ii. 21). All perception or imagination of determinate extensions of the wax will not exhaust the range of its possible extensions. But to grasp a thing's essence is, in part, to know all of its possible modes.[4] If I understand the wax at all, it is by virtue of the intellect and not sensation or imagination, which are limited to determinates. It is worth noting that the relevant difference is not that the intellect, as opposed to sensation or imagination, can encompass each individual member of $F_1 \ldots F_n$ one by one. Instead, what the intellect grasps is the nature or essence of the wax that includes $F_1 \ldots F_n$ in the sense that the idea of color, for instance, might be said to 'include' all determinate colors.

But what precisely is left over, after we have thought away all determinate modes of body? Descartes's language here might indicate that we are supposed to consider the wax apart from *any* of its properties, determinate or otherwise. On this reading, he is tacitly asserting a 'pincushion' model of inherence: properties require a substratum in which to inhere and which, on pain of regress, cannot itself be characterized by any properties. Pierre Gassendi seems to have read Descartes in this way:

Next you introduce the example of the wax, and you spend some time explaining that the so-called accidents of the wax are one thing, and the wax itself, the substance of the wax, is another. You say that in order to have a distinct perception of the wax itself or its substance we need only the mind or intellect, and not sensation or imagination. But the first point is just what everybody asserts, *viz.* that the concept of the wax or its substance can be abstracted from the concepts of its accidents. But does this really imply that the substance or nature of the wax is itself distinctly conceived? Besides the colour, the shape, the fact that it can melt, etc. we conceive that there is something which is the subject of the accidents and changes we observe; but what this subject is, or what its nature is, we do not know. This always eludes us . . . So I am amazed at how you can say that once the

---

[3] Thus, I do not see any reason to think that the wax argument has anything to do with the distinction between primary and secondary qualities, or the claim that the essence of bodies is extension. Instead, it is squarely focused on the determinate–determinable distinction.

[4] This is why Malebranche, for example, denies that we know the essence of the mind. We do not know all of the possible modes of minds, and hence we cannot know their essence.

forms have been stripped off like clothes, you perceive more perfectly and evidently what the wax is.   (AT vii. 271–2/CSM ii. 189–90)

From Gassendi's point of view, Descartes's particulars are too bare to be distinctly conceived. He scores these points, however, only by attributing to Descartes an ontology I shall show he simply does not hold.

In his reply to Gassendi, Descartes is too surly to be of much help to us. He writes,

Here, as frequently elsewhere, you merely show that you do not have an adequate understanding of what you are trying to criticize. I did not abstract the concept of the wax from the concepts of its accidents. Rather, I wanted to show how the substance of the wax is revealed by means of its accidents, and how a reflective and distinct perception of it (the sort of perception which you, O Flesh, seem never to have had) differs from the ordinary confused perception.   (AT vii. 359/CSM ii. 248)

What Descartes engaged in was not abstracting substance from accidents, or the concepts of the one from the other; on the contrary, the accidents themselves are, in some way we have yet to discover, the means by which the substance is known. It seems Gassendi misread the wax passage as implying the pincushion model.[5] It is much less clear what other model Descartes has available.

I think the precise nature of the substance–mode relation comes into focus only when we take account of Descartes's taxonomy of distinctions, as it appears in the Replies and the *Principles of Philosophy*.[6]

A *real* distinction obtains between *a* and *b* only when *a* and *b* can exist independently of one another. Only substances can strictly speaking be said to be really distinct, for 'by substance we can understand nothing other than a thing which exists in such a way as to depend on no other thing for its existence' (AT viiiA. 24/CSM i. 210). The epistemic criterion for the real distinction is simply our ability to clearly and distinctly understand one apart from the other (AT viiiA. 28/CSM i. 213). Descartes's claim that in the strict sense only God is a substance might cause unnecessary confusion. It is true that only God 'depends on nothing else' for his being, but there are two distinct senses of dependence at work. Since God must re-create the world at every moment, every created being depends on God. But the sense in which modes depend on substance is quite different—modes inhere in or modify a substance; they are not created by that substance. A substance is that which needs nothing else in

---

[5] Alternatively, one might take Gassendi's talk of the 'nature' of the wax to refer to its unknown explanatory structure. The difference between these readings closely parallels that between pincushion readings of Lockean substance and those that identify substance and real essence.

[6] See Descartes's replies to the First Objections (AT vii. 120 ff./CSM ii. 85 ff.) and his *Principles*, §§60–2 (AT viiiA. 28–30/CSM i. 213–14). In the *Principles*, Descartes admits that 'elsewhere [i.e., in the Replies] I did lump this type of distinction [the conceptual distinction] with the modal distinction . . . but that was not a suitable place for making a distinction between the two types; it was enough for my purposes to distinguish both from the real distinction' (AT viiiA. 30/CSM i. 215). See also the Letter to ***, 1645 or 1646 (AT iv. 349–50/CSMK 280).

which to inhere, and in this sense, we can say that God and creatures are alike substances.

A *modal* distinction obtains (1) between *a* and *b* when *a* is a mode (i.e., a property of a substance other than its essence) and *b* the substance it modifies, or (2) when *a* and *b* are both modes of a substance. Unlike the real distinction, (1) does not imply a common capacity for independent existence. Although we can clearly and distinctly conceive of substances apart from their modes, we cannot so conceive of modes apart from their substances. Once again, there is a potential source of confusion here. For after all, a substance must have *some* determinate mode or other at any given time; thus, a determinate quantity of extension will always have some determinate shape. The asymmetry between substance and mode thus cannot be captured simply by saying that modes require substances but not vice versa. But if we invoke the determinable–determinate distinction, we can make sense of the asymmetry. Modes are always determinate features of a substance, while the essence of a substance is a highest-order determinable.[7] The point, then, of saying that substances can be understood without their modes is that there is no one determinate mode the substance must possess. This makes it possible to clearly and distinctly conceive of the substance apart from the mode, while the mode cannot be so understood apart from the substance.[8] Similarly, in the case of (2), the modes cannot exist apart from the substance in which they inhere and therefore cannot be clearly and distinctly conceived apart from this substance.

Finally, a *conceptual* distinction obtains between *a* and *b* where *a* is a substance and *b* an attribute 'without which the substance is unintelligible,' or between two such attributes of a single substance (AT viiiA. 30/CSM i. 214–15).[9] This is a distinction only by courtesy, since things that are merely conceptually distinct 'are in no way distinct' (AT iv. 350). The difference lies entirely in the realm of thought, even if it has an objective foundation.[10] Conceptual distinctions arise from the fact that we can think of the same thing in two distinct ways, even though the thing itself remains a unity. Descartes offers a helpful example in the French version of the *Principles*:

in general all the attributes which cause us to have different thoughts concerning a single thing, such as the extension of the body and its property of being divided into several parts, do not differ from the body . . . or from each other, except in so far as we sometimes think confusedly of one without thinking of the other.[11]

Thus, in any given material substance there is no distinction between being extended and being divisible. Similarly, for a given mind to be a thinking thing

---

[7] I owe this interpretation to Jorge Secada; see Secada (2000).
[8] See AT viiiA. 29/CSM i. 213–14.
[9] For a detailed discussion of Descartes's theory of distinctions, see Secada (2000: 196 ff.).
[10] AT iv. 349–50.   [11] AT ixB. 53/CSM i. 215 n.; ellipsis in original.

and to be a substance is not for it to have two distinct properties but for it to be thought of in two different ways.[12]

This tripartite scheme lets us understand precisely how extension and its modes are related. Descartes disambiguates two very different senses in which extension can be said to be related to material substance. In one use, it is perfectly acceptable to call extension a mode of a substance. But all we can mean by this is that 'one and the same body, with its quantity unchanged, may be extended in many different ways' (AT viiiA. 31/CSM i. 215). That is, to call extension a mode of body is not to say that extension must itself modify some underlying substratum. Rather, the point is only that no *determinate* extension is essential to a body.

In another logical tone of voice, we can say that extension is the essence of body: 'Thought and extension can be regarded as constituting the natures of intelligent substance and corporeal substance; they must then be considered as *nothing else but* thinking substance itself and extended substance itself—that is, as mind and body' (AT viiiA. 30–1/CSM i. 215; my emphasis). This is a crucial aspect of Descartes's position, one that I think has not been sufficiently noticed by commentators. A substance *just is* its essence; this is signaled by the fact that extension stands to the extended substance in such a way that without the former the latter is unintelligible.[13] By contrast, to talk of extension as a mode is simply to recognize that any given body may have a variety of different sizes, shapes, and movements. There is nothing as it were lying behind the determinable quality, no cushion in which the pins stick. The distinction between extended substance and the determinable quality *extension* is merely conceptual.[14] This represents a departure from the Aristotelian model, which on at least one natural construal is a pincushion account, the cushion being prime matter.[15] Nor is this view peculiar to the *Principles*; in his conversation with Burman, for example, Descartes claims that 'all the attributes taken together are identical with the substance' (AT v. 155).

---

[12] As Secada puts it, 'the conceptual distinction applies between an existing real essence and its substantiality, existence, duration, order, and number. These are not different real properties, nor is any one a determinate mode of any other. . . . Rather, as second-order or derivative properties, they are the objects of "different thoughts" about "one same" property' (2000: 198).

[13] AT viiiA. 30/CSM i. 214.

[14] It has been suggested to me that, since there is also a conceptual distinction between a substance and its duration, my reading has the absurd consequence that there is nothing 'lying behind' the duration that endures through time. But this objection neglects the fact that, as a second-order property (to use Secada's language), substantiality, like duration and number, is simply another way of thinking about what is in reality a single thing. Duration in any actual substance is simply a way of thinking *of* that substance.

[15] The case is significantly more complex than this, of course, especially since prime matter doesn't exist; that is to say, it never exists as such. It exists only in so far as it is actual and hence not 'prime' after all. A further complication is that accidents do not, *tout court*, inhere in prime matter; they inhere in the substance, which is a form–matter composite. Nevertheless, the matter that makes up the substance (e.g., flesh and bone) is itself form relative to some lower level of matter, and so on, until we reach prime matter.

If we take seriously the notion that modes are simply ways essences have of existing, we can fit Descartes's claims in the *Principles* with his curt reply to Gassendi. Recall that Descartes takes himself in Meditation II to have described how 'the substance of the wax is revealed by means of its accidents.' The substance, which is identical with its essence, exists in various ways; once we figure out what those ways are, we are that much closer to grasping the essence, just as an awareness of multiple colors is at least a help in grasping the determinable *color*.

I have devoted a fair bit of space to arguing for this reading of Descartes partly because it will be important when we try to assess occasionalist arguments like the divine concursus argument and partly because it becomes an article of faith with later Cartesians, even those who disagree with him on many other crucial issues. Thus, Malebranche, for example, develops this position in *The Search After Truth*. After declaring that the mind's essence is thought, Malebranche adds: 'I warn only that by the word *thought*, I do not mean the soul's particular modifications, i.e., this or that thought, but rather substantial thought, thought capable of all sorts of modifications or thoughts, just as extension does not mean this or that extension, but extension capable of all sorts of modifications or figures' (*SAT* iii. i. 1: 198).[16] Modes are reserved for fully determinate qualities, while essences refer to determinables.[17]

With this background in place, we are in a position to look at its implications for causation. Now, it is not at all clear that Descartes in fact has an analysis of causation to offer.[18] What has often passed for such is what we might call the 'hierarchical principle.' In Meditation III, Descartes writes that 'there must be at least as much reality in the efficient and total cause as in the effect of that cause' (AT vii. 40/CSM ii. 28). As others have shown, this principle went through a variety of formulations in Descartes's work.[19] We should begin, however, by noting that it is not, on its face, an analysis of causation but a necessary condition on it.[20] Whatever it means, the principle does not explicate but merely constrains causation. But what precisely is this constraint?

---

[16] Compare Régis's definition of modes in his dictionary (*URF* 960): 'Le Mode n'est autre chose que le sujet, ou la substance ce même, entant qu'elle contient en substance toutes les façons et tous les états qui la peuvent diversifier.'

[17] This is not to say that *every* determinable is an essence. Color, for example, is a determinable, but not an essence.

[18] Jolley sums up most commentators' frustrations when he writes, 'Indeed, it is not too much to say that the whole theory of causality is unstable in Descartes's writings' (1990*b*: 40). For a very different analysis of Descartes on causation, see Flage and Bonnen (1997).

[19] See esp. Clatterbaugh (1999, ch. 2). I shall not repeat his thorough analysis here. See also Jolley (1990*b*: 41 ff.). Jolley uses the *Comments on a Certain Broadsheet* (AT viiiв. 359/CSM i. 304) to show that Descartes appeals to the likeness principle when he denies that sensory ideas cannot be adventitious. I think it more likely, however, that Descartes's innatist claim depends on the point that an idea of a properly sensorial quality bears no resemblance to the qualities of the object in the world. Thus, the content of that idea cannot be fully explained by those qualities.

[20] Thus, Schmaltz (2008: 84) points out that Descartes's causal axioms are compatible with conservationism, concurrentism, and occasionalism.

On one reading, it amounts to the claim that a cause must contain either formally or eminently any reality it transfers to its effect (i.e., if x causes y to become F, then either Fx, or x contains F in 'some grander way,' as Clatterbaugh puts it). But this talk of 'transfer of reality' is no less obscure than the principle itself, and seems merely metaphorical. I think the clearest formulation of the principle comes in the Replies to Hobbes. There, it is clear that if x causes y, then x is either at the same level of reality as y or higher, where levels of reality are understood in terms of the hierarchy *God–created substances–modes* (AT vii. 185/CSM ii. 130).[21] As Descartes puts it, each element in this hierarchy is 'more of a thing' than any falling below it. Note that the principle does not commit Descartes to the absurd view that a mode could itself serve as a 'total or efficient' cause of anything, even another mode. Whether and to what degree created substances can serve as causes will be the focus of the final chapter of Part I.

If this chapter, or indeed this part of the project, were intended as a complete account of Descartes's metaphysics, it would, of course, be woefully inadequate. I have not touched such issues as the ontological status of motion[22] and the individuation of material bodies.[23] What my purposes require, however, is not a thorough treatment of Cartesian metaphysics but an account of those regions of it that are directly relevant to the issues of causal power and law. And as we shall see, Descartes's novel version of the substance–mode ontology has further implications for his positions on force, law, and the causal powers of created beings.

[21] See also Axiom VIII of the Second Replies (AT vii. 166/CSM ii. 117), discussed below.
[22] The status of motion is particularly troubling. If a body is just a region of space, as Descartes's claim that space and extension are only conceptually distinct suggests, then it cannot move, whether motion is a real feature of the world or just God re-creating it in a different place. See Lennon (2007). Both Newton and Régis attack Descartes on the question of motion; see below, Ch. 13.
[23] I do address the latter, however, in Ott (2004*a*).

7

The Laws of Nature

Descartes is often credited with inventing the modern notion of a law of nature.[1] This is something of an overstatement. He might be among the first to apply that notion in the context of natural philosophy. But the concept of a law of nature was the descendant of the legal and moral concept and bears the marks of its lineage.

In April 1630, Descartes tells Mersenne that he plans, in a treatise on physics, to 'discuss a number of metaphysical topics and especially the following':

The mathematical truths which you call eternal have been laid down by God and depend on him entirely no less than the rest of his creatures. Indeed to say that these truths are independent of God is to talk of him as if he were Jupiter or Saturn and to subject him to the Styx and the Fates. Please do not hesitate to assert and proclaim everywhere that it is God who has laid down these laws in nature just as a king lays down laws in his kingdom.   (AT i. 145/CSMK 23)

Descartes here makes explicit his reliance on the metaphor of divine commands. The notion of a command grounds the typical sense of *lex* among the scholastics. As Aquinas puts it, 'Law is a rule and measure of acts, whereby man is induced to act or is restrained from acting; for law [*lex*] is derived from *ligare* [to bind],

---

[1] Much work in the history of ideas has been done on the metaphor of commands in the development of the concept of laws of nature; see especially Ruby (1986), Milton (1998), and Steinle (2002). The *locus classicus* of the view that the notion of the world as governed by law becomes possible only in the modern period is probably A. R. Hall (1954). For doubts, especially about Descartes's role, see Lehoux (2006). Lehoux suggests that most historians of science take the distinctive features of modern views of law to include the following: that the laws be descriptive rather than normative; specific rather than so general as to be useless; explanatory; and mathematical. Ruby points to the descriptive nature of modern concepts of law, along with the denial of a divine legislator and the view of nature as a set of regularities (1986: 350). One might wonder about Ruby's criteria: Newton, for instance, can hardly be said to deny the presence of a divine legislator (see Introduction, above); nor, given this, does it seem that Newton regards nature as *merely* a set of regularities, though of course all parties to the debate in the modern period agree that there are such regularities. Happily, my goal here is not to write the complete history of the concept of law but to see how that concept is developed and deployed in the modern period; I have no particular stake in whether it is utterly unprecedented. For my purposes, the important contrast is not whether there are laws in the sense of regularities that can be described in purely mathematical terms but whether these regularities are due to the underlying powers of objects or to the activity of a divine legislator. Even if the scholastics, for example, *could* have had a Newtonian concept of laws, they would have disagreed with Newton, Descartes, Malebranche, Berkeley, and others, on the origin and status of these laws.

because it binds one to act.'[2] Descartes's use of *lex* cannot be more than a metaphor, since the notion of subjects choosing to obey the law or not has no analog, either in physics or in mathematics. Nevertheless, Cartesian laws retain some features of their ancestors. They are arbitrary in the sense that they depend directly on the will of God.[3] (Whether God could have willed differently is a distinct issue I shall not address.) Unlike Ayers's mechanist, Descartes does not believe that the laws of nature follow from the essence of created substance, so that once God creates that substance, the laws follow of necessity. Instead, just as a king is not constrained by his subjects in decreeing his laws, God is not constrained by the nature of matter in his choice of laws.

Although the phrase *lex naturalis* most often occurs in scholastics like Aquinas in its moral sense—a prescriptive law derived from the commands of the creator—the notion that the world itself is governed by a divine being was far from foreign to them. To take just one example: quoting Cicero, who in turn quotes Aristotle, Aquinas writes that 'if we were to enter a well-ordered house, we would gather from the order manifested in the house the notion of a governor' (*ST* I q. 103 a. 1). God mediately directs the course of events by means of the powers and natures of the beings he has created.[4] '[T]he very notion of government of things in God, the ruler of the universe, has the nature of a law.'[5]

Late concurrentists such as Suárez also see the world as governed by divine law. Suárez claims that, once God has elected to create and conserve secondary causes,

> he concurs with them in their operations by an infallible law. If this law is taken absolutely—that is, without taking account of any particular and definite volition on God's part—it induces only a certain 'debt of connaturality' and not necessity; thus it is that God sometimes dispenses with this law by withholding his concurrence.   (*MD* 22. 4: 217–18)

This is not to say, however, that there is nothing new in Descartes's notion of law. Unlike Aquinas's God, Descartes's *im*mediately governs the world, since, as we shall see, Descartes's laws are aspects of God's will. This generates the further difference that for Descartes, certain of God's volitions—the laws of nature—govern the motion of bodies, whereas for Aquinas, God's volitions govern the creation of and concurrence with bodies and only through them their acts. Nor is Descartes simply repeating Suárez's view. Suárez says only that God acts in a lawlike way in concurring with secondary causes—his behavior is (again, leaving aside miracles) predictable, because immutable. But there is no

---

[2] *ST* I q. 90 a. 1.

[3] But as noted above, God's nature might well determine him to choose one set of laws over another.

[4] 'The natural necessity inherent in those beings which are determined to a particular course is a kind of impression from God, directing them to their end; just as the necessity whereby the arrow is moved so as to fly towards a certain point is an impression from the archer, and not from the arrow' (*ST* I q. 103 a. 1).　　　　　　　　　　　　　　　　　[5] *ST* I q. 91 a. 1.

suggestion of laws as particular explanatory principles.[6] Indeed, God's concurring with secondary causes is precisely what each and every non-miraculous event has in common, and so an appeal to the law by which God acts would be explanatorily vacuous.

Like the scholastic concurrentists, Descartes holds that 'the universal and primary cause—the general cause of all motions in the world' is 'no other than God himself' (AT viiiA. 62/CSM i. 240).[7] One might expect that Descartes would go on to substitute what Boyle calls 'the catholic affections of matter' for forms and occult qualities in the role of secondary causes.

But here Descartes turns, not to extension and its properties as secondary causes, but to laws. It is the 'laws or rules of nature, which are the secondary and particular causes of the various motions we see in particular bodies' (AT viiiA. 62/CSM i. 240). This is a decisive point in the history of mechanism.[8]

The concurrentists respond to the dilemma posed by God's ubiquitous creative act on one hand and the putative causal powers of objects on the other by adopting a bottom-up picture that accords causal powers to created beings and treats them as secondary efficient causes. But Descartes's ontological mechanism precludes this, and hence precludes course of nature mechanism. Recognizing this, Descartes locates secondary causation in the laws of nature. The ultimate sources of both causation and explanation, for natural phenomena at least, are thus not terrestrial but divine.

Descartes's laws of nature are really, of course, laws of motion, since this is the only way in which bodies interact or, more broadly, behave in ways that *could* be law-governed.[9] Motion is a mode of a body and is to be sharply distinguished from the act or force that puts a body in motion. In the *Regulae*, Descartes calls motion a 'simple nature' (AT x. 419/CSM i. 45); this terminology is dropped in the *Principles*, although Descartes continues to refer to motion as 'simple in itself.' There, he defines motion as 'the transfer of one piece of matter, or one body, from the vicinity of the other bodies which are in immediate contact with it, and which are regarded as being at rest, to the vicinity of other bodies' (AT viiiA. 54/CSM i. 233).[10] Note that Descartes speaks not of *actus* or *vis* but of *translatio*: motion is not to be identified with action, force, or power, but with a mode of the thing moved.

---

[6] For more on this point, see Milton (1998).

[7] It is worth noting that while the scholastics would take *motus* in a very broad sense to include all changes, Descartes is thinking of local motion. But since all changes on his view are due to local motion, the difference is not as great as first appears.

[8] By contrast, Clarke (1989: 104 ff.) and Hattab (2007) discuss primary and secondary causes without seeing the force of Descartes's casting laws of nature as secondary causes. See below.

[9] This awkward phrase is designed to accommodate such behavior as is described by the law of inertia, which is not, except in a limiting case, a law about how bodies *interact*.

[10] The French version reads thus: motion 'est le transport d'une partie de la matière, ou d'un corps du voisinage de ceux qui le touchent immédiatement, et que nous considérons comme en repos, dans le voisinage de quelques autres.'

Although the laws governing motion swing entirely free of the bodies whose behavior they prescribe, and hence cannot be read off from them, this is not to say that they are unknowable. Descartes maintains that much of God's nature and purposes must remain hidden from us; this is part of his case against the use of final causes in physics. But there is one feature of God we can be sure of: his immutability. 'We understand that God's perfection involves not only his being immutable in himself, but also his operating in a manner that is always utterly constant and immutable' (AT viiiA. 61/CSM i. 240). And it is this feature which both grounds the status of the laws as *laws* as opposed to mere regularities and allows us to deduce them a priori: 'From God's immutability we can also know certain laws or rules of nature' (AT viiiA. 62/CSM i. 240).

The laws of nature are derived, both epistemically and metaphysically, from God's nature alone. By casting these laws in the role traditionally assigned to the powers of bodies, Descartes turns the whole of natural philosophy on its head. Rather than merely providing a handy way to talk about the orderliness of nature, laws now assume a central role in natural philosophy, a role they could never have had on the bottom-up picture of the scholastics.

As we might expect, this achievement in the *Principles* follows a period of flailing about, in which Descartes's views are considerably more opaque and ambiguous. As I have indicated, I think Descartes's position undergoes a transformation: early texts like *Le Monde* and the *Discours* are ambiguous, and incline toward conservationism or concurrentism. It is only in the *Principles* that Descartes self-consciously hijacks the language of concurrentism and bends it to his own very different purposes.

To see how and why this development happens, we can begin by recalling the chief role the concurrentists assign to secondary causes. On Aquinas's view, for example, secondary causes are required to explain the diversity and mutability of the natural world. Aquinas writes, 'if God works alone in all things, then, since God is not changed through working in various things, no diversity will follow among the effects through the diversity of the things in which God works.' This is 'evidently false to the senses' (*SCG* 69: 125). It is only through creatures that God's activity, which is as immutable as himself, is diversified into the various forms we see around us. God, being immutable, cannot on his own suffice to explain the diversity of things we see around us. How exactly does Descartes propose to account for this diversity?

In his earlier work, Descartes adverts not to secondary causes but to the modes of bodies. Since God continues to preserve the world in the same way he created it, Descartes writes, 'there must be many changes in its parts which cannot, it seems to me, properly be attributed to the action of God (because that never changes)' (AT xi. 37/CSM i. 93). This is just Aquinas's point: an immutable God cannot be responsible, immediately and directly, for everything that happens. *Le Monde* goes on to divide up the causal work: God is the cause of rectilinear motion and motion as such, but the diverse modes of matter divert this motion into its other forms. It is important to be clear just how different this is from concurrentism. On that view, God and creatures are jointly responsible for any given effect, in the sense that God works through those creatures to accomplish his ends. On the view suggested by *Le Monde*, however, God is responsible for some features of the effect, and bodies others. This is a quite different picture. Descartes's view in these early texts is really a form of conservationism, not concurrentism: there is no suggestion that God 'works through' bodies to achieve effects.[11]

In both *Le Monde* and *Discours*, Descartes claims that, even if God initially created a mass of bodies in motion devoid of order or proportion, the laws of nature would suffice to educe order from chaos. How this happens is, in each case, importantly different. As we have just seen, *Le Monde* requires that bodies bring about some changes in motion on their own. In the *Discours*, Descartes's God 'variously and randomly agitated the different parts of this matter' to produce, in the first instance, mere chaos. The emergence of order requires only that God 'lend his regular concurrence to nature, leaving it to act according to the laws he established' (AT vi. 42/CSM i. 132). Again, this sounds more like conservationism than anything; nature acts autonomously, though according to laws. God 'established the laws of nature and then lent his concurrence to enable nature to operate as it normally does' (AT vi. 45/CSM i. 133). 'Concurrence' here might simply mean God's continual existential support, or it might mean something stronger. But the stronger reading is belied by Descartes's talk of nature 'operating' as it normally does; again, there is no hint here of God working through nature or bodies as through his instruments. Whether concurrentism or conservationism is the appropriate term, *Le Monde* and the *Discours* clearly assign bodies a key role in explaining the diversity of motions and the orderliness of nature.

But some dozen or so years after *Le Monde*, Descartes substantially alters his view. Now, *the laws of nature* are 'the secondary and particular causes of the various motions we see in particular bodies' (AT viiiᴀ. 62/CSM i. 240). The particular modes of bodies are nowhere to be found in this description. Whatever his earlier views, the Descartes of the *Principles* casts laws and not modes of bodies as secondary causes. But are they up to the task of explaining the diversity of motion? Just as Aquinas's secondary causes 'particularize and determine the action of the first agent' (*SCG* 66: 119), Descartes's secondary causes must be responsible for the precise form in which motion is manifested in the world. But the laws are no more mutable than God himself.

Descartes seems to have a serious problem on his hands. It is not just that it is mysterious how an immutable God could create anything in the first place; many theists, Aquinas included, are faced with that difficulty. The interesting

---

[11] For a genuine concurrentist in the mechanist mold, we have to look to Régis; see below, Ch. 13.

problem, I think, is why Descartes would find it anything but self-contradictory to announce that the diversity of motion in the world can be accounted for by laws that do not themselves change. The trick is turned by noting that 'God imparted various motions to the parts of matter when he first created them' (AT viiiA. 61/CSM i. 240). God, being immutable, then goes on to preserve exactly the same quantity of motion in the world and distribute it according to the laws we shall explore below. So an initial variety of motions, plus an immutable God, can suffice to explain the bewildering variety of motions exhibited by objects in our experience.[12] Now, Descartes is still left with the problem of reconciling God's initial creation of various motions with his immutability. But as I have said, Descartes has this in common with every other theist who goes in for an immutable God. The vital point is that, in contrast to the accounts of *Le Monde* and the *Discours*, the diversity of motion is explained, not by the operation of bodies, but simply by the original diversity of motion God produced at the time of creation, plus God's continued law-governed activity.

To sum up what we have established so far: Descartes's laws of nature serve as a replacement for scholastic secondary causes in that they (help to) account for the diversity of motion and serve as causes of this motion. In addition to God's creation of motion, they are supplementary principles of natural change. Now, it is not easy to understand how laws of nature, construed as propositions, could serve as causes in any sense, though Malebranche, too, speaks of laws as causes.[13] Being neither substances nor modes, they are simply not of the right category to exist independently of God, even to the degree that created beings exist. As we have seen, scholastic secondary causes are genuine collaborators in the production of their effects; they are, that is, a kind of efficient cause. Descartes's laws seem, however, unable to play this role. One alternative here is to take Descartes's use of 'secondary causes' as a self-conscious attempt to bend a bit of scholastic jargon to his own purposes. It would hardly be the first time he did this.[14] It is important to recall that Descartes's initial hope for the *Principles* was that it would be taken

---

[12] For a different view on this issue, see Schmaltz (2008: 126), who argues that, since God's activities must have a 'constant effect,' God himself cannot be the sole cause of bodily events, for in that case, they would be similarly changeless. This is the centerpiece of his conservationist reading. By contrast, I think the initial variety produced by God allows his 'constant effect' to be diversified.

[13] '[God] also willed certain laws according to which motion is communicated upon the collision of bodies; and because these laws are efficacious, they act, whereas bodies cannot act. There are therefore no forces, powers, or true causes in the material, sensible world' (*SAT* iii. ii. 6: 449). For his part, Schmaltz (2008: 115; see 88) holds that the efficacy of Cartesian laws is to be cashed out in terms of the bodily forces and dispositions that are responsible for changes in the distribution of motion. One might worry, however, that laws, on this view, are causes only in the most attenuated sense: they are at best a shorthand for regularities arising from the forces and inclinations of bodies. Again, I think this is in conflict with the top-down nature of Descartes's views on laws. (I discuss this problem at greater length in the context of Malebranche's view below.)

[14] Consider, for example, Descartes's use of formal and objective reality in Meditation III and the First Replies.

up by the schools as a textbook, and so he has a motive for playing down, so far as possible, his departures from the professors. But we can go further. If laws are volitional contents, they at least have a role to play in the effects of the divine will, even if, strictly speaking, it is that will itself that is the genuine efficient cause. Moreover, the fact that they play the other key role assigned to secondary causes—explaining the diversity of effects that ultimately have their source in an immutable being—might well be enough to justify Descartes's use of the scholastic terminology. Still, we should not allow this terminological similarity to obscure the radical flavor of Descartes's departure.

I now propose to turn to the *Principles'* treatment of the laws themselves. My discussion can be brief, since others have covered this same terrain.[15] Contrasting the derivations of the laws with those in *Le Monde* will let us see in more detail how Descartes's view develops. More important, it sets the stage for the analysis of force that is to follow.

> *Law 1.* Each and every thing, in so far as it can, always continues in the same state; and thus what is once in motion always continues to move. (*Principles*, II. 37, AT viiiA. 62/CSM i. 240)

The scholastic opinion that motion of itself tends to diminish is nonsensical, since nothing can 'by its own nature tend towards its opposite' (AT viiiA. 62/CSM i. 241; Descartes makes the same point in *Le Monde*, AT xi. 40/CSM i. 94). What encourages this false opinion is the fact that the motion of a body is often interfered with by the collision of other, insensibly small bodies. Even so, had the scholastics seen that motion is a mode, they would have accepted Law 1. For they endorsed an analog of Law 1 with regard to other qualities such as shape and size, and even rest; but by defining motion as they did—'the actuality of a potential being in so far as it is potential'—they were unable to see that motion, just as much as shape or size, is a mode (AT xi. 39/CSM i. 93–4).

It is tempting to think that when Descartes says that each thing 'in so far as it can' ('quantum in se est', 'autant ce qu'il peut') persists in the same state, he means that the body itself is the source of its own motion.[16] This would flip Descartes's top-down nomology on its head, and make a quality of bodies responsible for at least some of the events in which those bodies figure. It would also make it obscure how this law could follow from God's immutability. But we cannot read Descartes in this way. It should already be

---

[15] See esp. Garber (1992); see Stein (1990) for criticism of Descartes.

[16] This is a key element in Schmaltz's reading; see Schmaltz (2008: 116 ff.). On his view, *Principles* II. 43 appeals to the tendency of objects to endure in the same state. This allows him to (quite neatly) analyze forces: the force of a given body is its tendency to persist in the same state, its 'strength of duration' (119). Bodies enduring for the same quantity of time can nevertheless have distinct tendencies to endure in their present states, and thus differ in terms of force. Schmaltz thus parries Gabbey's 'no existential variation' objection (see Gabbey 1980: 237). (I am indebted to correspondence with Schmaltz on these issues.)

clear that the reason something persists in the same state is not because of any features it has but because of God's uniform operation as the source of motion. To talk of a body continuing along in a given direction and speed 'quantum in se est' is only a way of saying that the body will do so, in virtue of God's nature and will, unless and until God brings some other body in contact with it to impede or divert its motion. That Descartes explicitly holds the position I am attributing to him can be seen from his denial of inertia construed as a real property of beings. In a letter to Mersenne (December 1638), Descartes writes, 'I don't recognize any natural inertia or sluggishness in bodies' (AT ii. 466–7/CSMK 131).[17] It is difficult to suppose, then, that his talk of bodies persisting in the same state 'quantum in se est' is meant to attribute just such a quality to them.[18]

> *Law 2.* All motion is in itself rectilinear; and hence any body moving in a
>   circle always tends to move away from the center of the circle which
>   it describes. (*Principles*, ii. 39, AT viiiA. 53/CSM i. 241)[19]

Like Law 1, this law is supposed to flow from God's immutability. In both the *Principles* and *Le Monde*, the thought seems to be that 'God always preserves . . . motion in the precise form in which it is occurring at the very moment when he preserves it, without taking any account of the motion which was occurring a little while earlier' (AT viiiA. 63–4/CSM i. 242; cf. AT xi. 44/CSM i. 96). Thus, *ceteris paribus*, the body moves in a straight line. This law itself is not so different from Law 1, and may be simply a consequence of it. What counts for us is not the derivation of this law itself but rather the support it offers for my developmental thesis.

In *Le Monde*, Descartes argues that 'God alone is the author of all the motions in the world in so far as they exist and in so far as they are rectilinear; but it is the various dispositions of matter which render them irregular and curved' (AT xi. 46/CSM i. 97). Acting in the same way, God 'always produces the same effect' (AT xi. 43/CSM i. 96). This prevents God from being the complete explanation of a body's moving in a curve. On this picture, bodies must be genuinely causally active in a more robust way than that envisioned even by concurrentism. It is not that God works through bodies to produce motion in a curve, as the concurrentist would have it; rather, God produces rectilinear motion which then gets deflected by bodies in various directions.

It is striking that this line of thought disappears altogether in the *Principles*. There is no suggestion that bodies are directly responsible for non-rectilinear motion. Instead, Descartes simply repeats his point about how God conserves motion and remarks that Law 2 has the same basis as Law 1.

---

[17] Descartes's target here seems to be Kepler's notion of inertia as a force present in bodies.

[18] I do not mean to deny, of course, that inertia truly characterizes the behavior of bodies. The point is that, on Descartes's view, it does so in virtue of God's activity, not any intrinsic feature of the bodies themselves.   [19] In *Le Monde*, Laws 2 and 3 switch places.

> *Law 3.* If a body collides with another body that is stronger than itself, it loses none of its motion; but if it collides with a weaker body, it loses a quantity of motion equal to that which it imparts to the other body. (*Principles*, II. 40, AT viiiA. 65/CSM i. 242)

The intriguing question with regard to this law is the ontology of force that Descartes seems to help himself to. One gets the impression that in the 'impact contest,' as Garber calls it, the body with the greater amount of force wins. But what sense can be made of 'strong' and 'weak,' when extension is exhausted by size, shape, and motion? The obvious answer is that these are mere shorthand; to say that a body is stronger is to say something about these other properties, and ultimately something about God's will as directed at them. (I shall have more to say on this in the next section.)

My general strategy will be to suggest that much of what Descartes says in these laws is also shorthand that will have to be cashed out in quite different terms.[20] As a first example, take Descartes's talk of a body 'imparting' motion to another. As a mode of a body, motion cannot literally be imparted to another body, for this would reintroduce real qualities. That is, if motion were literally transferred, it would have to migrate from one body to another, and this is flatly inconsistent with Descartes's rejection of real qualities (see his Letter to Elizabeth, May 21, 1643, AT iii. 667/CSMK 219).[21] What he means is simply that a body with a given motion, on encountering another, acquires a different quantity of motion than it had before. In general, wherever Descartes's language conflicts with his explicit commitments, we should look for an innocuous paraphrase of that language.

What is significant in all of this is the way in which the dependence of the laws of nature on God seems to rule out the causal efficacy of created substances. This is a difficult point, and we must be clear precisely what sense of dependence is at issue. Indeed, I shall argue below that Régis is able to make sense of the dependence of the actions of created beings on God while preserving their causal powers. This move is not available to Descartes, however, since he rejects the bottom-up picture. And having done this, he can no longer accord causal powers

---

[20] One might object that this strategy is not fair pool, given that it requires reading away much of what Descartes says. On the contrary: given the many prima facie tensions in Descartes's view, any interpretation that makes him at all consistent will have to do some amount of 'reading away.' Second, there are principled reasons for taking some passages as definitive and others not. These reasons are of course contextual in nature and always specific to the passages in question. In general, though, we can say that, when faced with conflicting passages, we ought to determine which (*a*) fits best with other things Descartes unambiguously says and (*b*) can be expected, because of their dialectical purpose and context, to reveal Descartes speaking strictly rather than loosely.

[21] Schmaltz (2008: 113) argues that Descartes's containment axiom—the view that the cause must contain formally or eminently its effect, discussed above—does not entail that modes can migrate from one substance to another. Qualitative, rather than numerical, identity is enough. And as Schmaltz notes, Descartes himself explicitly rejects the migration theory in a letter to More (AT v. 405/CSMK 382).

to creatures. Put roughly, if created substances had causal powers, they would then have some role in determining the course of events; that is, any complete *causal* explanation of a sublunary event would have to appeal not just to the laws that God has arbitrarily decreed but to the properties of the substances involved. But created substances have no such role, on Descartes's view.[22] Had God created not extension but shmextension, the laws of nature might have been precisely the same.

[22] This is not to say, of course, that a true description of an event in Descartes's physics will not mention the determinate modes of bodies (just as it would in Malebranche's physics). It is just to say that those modes are not efficient causes of those events. See the following chapter.

8

Force

Although I think the strong top-down nature of Descartes's account is undeniable, it does seem, on its face, to conflict with some of Descartes's claims about *vis* ('force' or 'power'). It is uncontroversial that Descartes often speaks of *vis* as if it were a novel property of extension, alongside size, shape, and movement, and this is surprising, given his inveterate opposition to scholastic powers.

For example, writing to Mersenne in 1640 about the work of a certain Father Lacombe, Descartes has this to say:

He is right in saying that it is a big mistake to accept the principle that no body moves of itself. For it is certain that a body, once it has begun to move, has in itself for that reason alone the power [*la force*] to continue to move, just as, once it is stationary in a certain place, it has for that reason alone the power to continue to remain there. (AT iii. 213/CSMK 155)

Now, this cannot, and does not, mean that body contains within itself a principle of motion such that it could begin to move on its own. What Descartes says here is fully consistent with his statement to More in 1649: 'I agree that "if matter is left to itself and receives no impulse from anywhere" it will remain entirely still' (AT v. 404/CSMK 381). But the problem remains: how can bodies have a *vis* or a power to behave in certain ways?

A variety of increasingly ingenious methods have been worked out to reconcile force with Descartes's ontological mechanism.[1] I think the answer here is rather simpler than has previously been thought. Roughly, force talk is to be analyzed away in favor of law-of-nature talk. To make this case, let us return to *Le Monde* and examine the remainder of the passage I quoted above (Chapter 5).

For you may posit 'fire' and 'heat' in the wood, and make it burn as much as you please: but if you do not suppose in addition that some of its parts move about and detach themselves from their neighbours, I cannot imagine it undergoing any alteration or change. On the other hand, if you take away the 'fire,' take away the 'heat,' and keep the wood from 'burning'; then, provided only that you grant me there is some power which puts its finer parts into violent motion and separates them from the coarser parts,

---

[1] The most convincing of these, to my mind, is Schmaltz's (2008). For arguments that Cartesian bodies do *not* have force or power, see Hatfield (1998). Hatfield does a fine job of explaining Descartes's use of the notion in his more scientific works and I see no need to repeat his arguments here.

I consider that this power alone will be able to bring about all the same changes that we observe in the wood when it burns.   (AT xi. 7/CSM i. 83)

What exactly is this power that sets the parts into motion? To answer this, we need to look to the *Principles*, where Descartes offers to explain the nature of the power bodies have to act on each other. (Note that this section comes as an addendum to the discussion of the three laws of nature, and is not a part of it.) This power is identified neither with a real quality nor with the properties of extension Descartes adverts to in *Le Monde*; instead, 'This power consists simply in the fact that everything tends, so far as it can, to persist in the same state, as laid down by our first law' (AT viiiA. 66/CSM i. 243).

As I have argued, the first law's talk of a body persisting in the same state 'quantum in se est' is shorthand for God's tendency to conserve a body in the same state, *ceteris paribus*. Putting these passages together, we find that talking of power is a disguised way of talking about the inevitable results of God's unchanging nature. And since power in this sense is not a quality, real or otherwise, it is strictly speaking a category mistake to attribute a power to bodies at all. This is what we should expect, since the primary concept for Descartes is that of *lex*, not *vis*.[2]

Other pressures pushing Descartes toward an eliminative reduction of *vis* in the natural world derive from his rejection of real relations. *Vis*, construed as a real, additional feature of bodies, cannot be fitted into Descartes's substance–mode ontology. For to say that x has the *vis*, taken in this way, to continue in the same direction is at least to say that x stands in certain relations to possible, and perhaps never actual, states of affairs. But this sort of thing is precisely what draws Descartes's fire, in the 'little souls' argument. To attribute *vis* to bodies is to court occultism. The only sense to be made of *vis* in the context of bodies is to take it as an abbreviation for a complex set of tendencies grounded in God's will. There is no way to reduce *vis* to the other properties of bodies. By way of closing this chapter, it is worth examining a clever attempt to do just this.[3]

In discussing Descartes's means of replacing the scholastic entities such as substantial forms, Andrew Pessin writes,

bodies will be 'empowered,' hence have 'forces' (to persist in or resist motion, or affect other bodies' motions) not because of any 'substantial forms' but only insofar as they fulfill the antecedents of the laws, by virtue of their modes; so 'force' in bodies is not something *added to* extension and its modes, but reduces *to* extension and its modes.[4]

In support of this, Pessin cites *Principles* ii. 43, discussed above. But there we saw Descartes analyzing talk of force in bodies as a shorthand for talking about the laws of nature. So force cannot be reduced to the modes of extension

---

[2] Thus, my reading of Descartes on *vis* is consistent with that of Garber (1992: 298).

[3] For other attempts, see Gueroult (1980) and Gabbey (1980). I think the best treatment is to be found in Hatfield (1998).   [4] Pessin (2003: 40; cf. 48).

simply because force, as a logical fiction, is not of the right ontological type to be reduced to anything. Now, Pessin is right when he says that bodies fulfill the antecedents of laws in virtue of their modes. When God applies the laws of nature to bodies, he at least must look at their current state to decide which law to apply. But this in no way makes bodies causally active. To see this, consider that even Malebranche's God does precisely the same thing. The occasionalist view treats the states of bodies as occasions for God's activity; this hardly makes them causally active, any more than the notes on a page of manuscript cause the violinist to play.

If I am right, then attributing force to bodies is a convenient shorthand for talking about how God is liable to move them around. Descartes has provided an eliminative reduction of *vis* in the natural world; but this is not the same as providing a reduction of *vis tout court*. He has instead simply located *vis* somewhere else. For all the contempt Descartes displays toward scholastic notions of power, he still believes that minds are substantial forms capable of initiating motion on their own. This is just the flip side of the little souls argument. If bodies cannot be intrinsically directed at non-actual states of affairs, as power ascription requires, Cartesian minds (including God) certainly can. In short, most of the features examined above that made the attribution of powers to bodies so implausible to Descartes are perfectly acceptable to him when attributed to minds.

9

Occasionalism

Descartes's positions on laws and force seem to push him toward some version of occasionalism. But seemingly equipollent considerations tell against this. Much of the debate in the last twenty years has focused on precisely this question. But recently a new interpretation has emerged, promising to effect a reconciliation: concurrentism.[1] On this reading, Descartes accommodates the causal contributions of bodies in much the same way Aquinas does: by treating them as secondary efficient causes, instruments through which God does his work.

As we have seen, there are in fact four competing interpretations: conservationism, concurrentism, and two kinds of occasionalism: 'limited occasionalism' and 'thoroughgoing occasionalism.' Limited occasionalism asserts that bodies have no causal powers and that God is the only cause of body–body interactions. It is neutral with regard to the role minds might play in the world. A thoroughgoing occasionalism, then, is one according to which God is the only true cause full stop; neither minds nor bodies are possessed of causal powers. To complicate matters, any assessment of Descartes's position must be sensitive to the distinctions between these three questions:

1. Does Descartes self-consciously hold the view?
2. Is Descartes implicitly committed to it?
3. Does Descartes make some pronouncements that entail the view, even if other features of his position conflict with it?

I shall argue that concurrentism deserves a 'no' on all counts, if we are looking at Descartes's mature view, embodied in the *Meditations* and *The Principles of Philosophy*. While the *Discours* might contain hints of a concurrentist view, *Le Monde* is clearly conservationist. I think these early efforts are best understood as Descartes struggling to arrive at a view consistent with his other commitments and hence experimenting with various permutations of these doctrines. But I shall continue to argue that Descartes ultimately changes his mind and turns course toward occasionalism in his mature works.

---

[1] See, e.g., Clatterbaugh (1995, 1999) and Pessin (2003). I do not wish to underestimate the force of Schmaltz's (2008) conservationist reading; see above.

Limited occasionalism, then, emerges as a strong contender. My main interest is in seeing just how the *Principles* conception of laws leads to the denial of body–body causation. But an argument for thoroughgoing occasionalism is also available—the so-called 'divine concursus' argument, offered by later Cartesians such as La Forge and Malebranche. If God re-creates the world at each moment, there seems to be little left for bodies to contribute to the future states of the universe. The difficulty created by this argument, of course, is that it then becomes difficult to stop at limited occasionalism: if successful, it would rule out not just bodies as causes but minds as well. And Descartes's commitment to the causal powers of minds is unwavering at all points in his career.

Where limited occasionalism is concerned, then, I believe that (1)–(3) all merit a 'yes.' As to thoroughgoing occasionalism, however, (1) must be answered 'no.' Nevertheless, I believe that Descartes is indeed implicitly committed to it, even in the face of countervailing considerations. Thus, in the final section below, I turn to the issue of mind–body causation, and argue that any attempt to reconcile it with limited occasionalism must fail. In some ways, this is a disappointing result; Descartes emerges as a figure drawn to contradictory positions. But this hardly makes him unique in the panoply of great philosophers.

9.1 THE CONCURRENTIST READING

The main advantage of a concurrentist reading is that it allows us to reconcile Descartes's claim that God is the cause of everything and his use of causal language when discussing bodies in scientific contexts. As Clatterbaugh points out, there are many texts in which Descartes speaks of bodies both as causes and as vital elements in scientific explanations. In the *Optics*, for instance, Descartes speaks of his explanation of refraction as the *cause* of those refractions (AT vi. 104/CSM i. 163). Such passages could easily be multiplied. If concurrentism is right, created beings can be genuine causes, while still requiring God's activity.

Clatterbaugh notes that Descartes sometimes treats causes as things and sometimes as premises. He finds this usage puzzling, but it seems fairly unproblematic if we take cause-as-premise simply to mean premise-that-states-a-cause. Talking of causes as premises, then, is just a way of talking about what those premises refer to.[2] In any event, Clatterbaugh convincingly argues that Descartes takes events in the natural world to be in principle deducible from an initial set of premises. Writing to Plempius in 1637, Descartes describes the principles from which his ideal science would be drawn:

Sizes, shapes, positions and motions are my *formal* object (in philosophers' jargon), and the physical objects which I explain are my *material* object. The principles or premises

---

[2] Flage and Bonnen (1997) take this to support their reading of Descartes as analyzing causation as *formal* causation.

from which I derive these conclusions are only the axioms on which geometers base their demonstrations: for instance, 'the whole is greater than the part' . . . but they are not abstracted from all sensible matter, as in geometry, but applied to various observational data which are known by the senses and indubitable.   (AT i. 476/CSMK 77)

These early remarks are hardly consistent with the mature view of the *Principles*. The idea that Cartesian physics requires only geometrical premises is simply untenable, in light of the derivation of the laws of nature this later work offers. Still, it seems fair to say, with Clatterbaugh, that explanations of phenomena would have to include 'both God-premises and created thing-premises.'[3] Even the mature physics requires statements not just of the laws of nature but of the positions and motions of bodies. While true, this is not enough to secure the concurrentist conclusion. Even a thoroughgoing occasionalist like Malebranche would agree that created thing-premises have a role to play. The question is not, can we deduce the course of events without paying any attention to the objects that figure in those events (this is absurd), but whether those objects play any role in producing them. An occasional cause is neither necessary nor sufficient for its effect, and so it is not a cause after all.

So the easy route to concurrentism is just not available. Even if Clatterbaugh is right, and explanations of phenomena require created thing-premises, this does not yet show them to be causally active, even as secondary causes. In some texts, such as those Clatterbaugh adduces from the *Meteorology*, Descartes links causation and explanation, so that to explain a thing is to give its cause. But there is no reason to think that Descartes is particularly careful about his causal language when the context is not that of first philosophy.[4] What is more to the point, given the relative novelty of mechanist body–body occasionalism, it would be unreasonable to expect Descartes to sever explanation from causation in the explicit way Malebranche does. By contrast, the concurrentist tradition and its language of primary and secondary causes *was* well known, which makes it all the more puzzling why Descartes would not use those terms in anything like the standard way if he were in fact a concurrentist.

---

[3] (1999: 62).

[4] It is worth noting that all parties to this debate will have to read away a certain amount of textual evidence. For example, in a letter to Elizabeth, Descartes writes that 'God is the universal cause of everything in such a way as to be also the total cause of everything; and so nothing can happen without his will' (AT iv. 314/CSMK 272). Although reading Descartes as a body–body occasionalist, Garber does not take this passage as decisive, since he thinks it is meant 'as consolation, not metaphysics' Garber (2001: 205). Why these two should be mutually exclusive is obscure to me. Why not read Descartes simply as consoling Elizabeth by pointing to a feature of his metaphysical view? For his part, Pessin argues that by using 'total cause' here, Descartes means only that God is the principal or most important cause, and that created beings have a role to play as well. But Pessin's argument for this reading is strained. He claims that 'the total cause [in some texts] seems to be not the complete set of causes, but rather that member of the set that in some sense is superordinate' (2003: 43). But his sole example of such a use is precisely the text from the letter to Elizabeth we are examining. So he simply begs the question; to substantiate his point, he would need to offer *other* uses of 'total cause' to mean 'most important from among a set of causes.'

If this easy argument is not available, we might look to other textual evidence. Arguing against the conservationist (or 'mediate concurrentist') reading, Andrew Pessin adduces the fifth replies, in which Descartes is challenged to explain just how he understands God's continual activity.[5] Descartes writes,

When you deny that in order to be kept in existence we need the continual action of the original cause, you are disputing something which all metaphysicians affirm as a manifest truth—although the uneducated often fail to think of it because they pay attention only to the causes of *coming into being* and not the causes of *being itself*. Thus an architect is the cause of a house and a father of his child only in the sense of being causes of their coming into being; and hence, once the work is completed, it can remain in existence quite apart from the 'cause' in this sense. But the sun is the cause of the light which it emits, and God is the cause of created things, not just in the sense that they are causes of the *coming* into being of these things, but also in the sense that they are the causes of their *being*; and hence they must always continue to act on the effect in the same way in order to keep it in existence. (AT vii. 369/CSM ii. 254–5)

Pessin's point here is that God's activity is not limited to preserving objects in existence—as mere conservationism has it—but also includes bringing about all changes in the material world. Now, as we have seen, there is a sense in which the concurrentist can say this. God, in concurring with creatures, is part of the total cause (i.e., the total set of individually necessary and jointly sufficient conditions) of any effect. But in fact Descartes's text says something much stronger. God is the cause of both the coming into being and the being itself of any creature. There is no suggestion here that created beings cooperate in any way. It is true that this text is not definitive evidence for limited occasionalism because of its context. Descartes is arguing for his principle that God must re-create the world at every moment, and so he might understandably neglect the role of creatures as secondary causes. But all this means is that this passage, like the others Pessin cites, is simply neutral between concurrentism and occasionalism.[6] To decide among the competing interpretations, we must look elsewhere. But where?

I have already argued for a developmental thesis: Descartes begins with some form of conservationism and then changes his mind. The obvious question is, why? I suspect that in the interval between these early works and the *Principles*, Descartes came to see the tension between his derivation of the laws of nature and the suggestion that bodies might operate on their own, dictating the behavior of matter without God doing any more than providing his 'regular concurrence.' Recall the *Discours* claim that God leaves matter 'to act according to the laws he established' (AT vi. 42/CSM i. 132). This makes sense on a bottom-up reading of those laws; but it is precisely this reading that the *Principles*' derivation of those laws rules out.

[5] See Pessin (2003: 37).
[6] Note that I have dealt with another of Pessin's arguments above, in Ch. 8.

All of this is in addition to the difficulties the early Descartes faces in making sense of *vis*. The earlier writings represent an awkward developmental stage, in which Descartes is poised between the concurrentism or conservationism of his intellectual antecedents and the body–body occasionalism of his mature period. What pushes him along in this development, I believe, is a desire to endorse the implications of the top-down picture he was in the process of inventing. The crucial feature of this view, and the source of the most persuasive argument against the concurrentist reading, is its conception of the laws of nature, to which we now turn.

9.2 THE ARGUMENT FROM LAWS OF NATURE

Despite its appeal, the concurrentist reading of Descartes's mature view cannot be right. The chief problem lies in the attribution to Descartes of the distinction between primary (God) and secondary (bodies) efficient causes, which is the linchpin of the concurrentist view. Descartes simply does not have this distinction; or rather, he has the wrong one. Secondary causes for Descartes are not bodies but laws. Without a division of labor between God and created beings, Descartes lacks the materials with which to build a concurrentist position. And given that he obviously was familiar with scholastic concurrentism, and indeed welcomes agreement with the scholastics where he can find it, it is mysterious that Descartes never puts his position in their terms.

But the decisive point against the concurrentist reading, and against any reading that attributes causal powers to bodies, is its inability to make sense of the laws of nature and their derivation.

What can it mean to say that the laws of nature flow from God's will, when one has endorsed a top-down picture of nomological necessity, other than that God has elected to move bodies about in certain regular ways? If the laws of nature supervened on created beings and their properties, as on the bottom-up view, we could easily make sense of according causal powers to bodies. For God to be responsible for the laws of nature would simply be for him to have chosen which bodies to create and conserve, and with which powers to concur. But when there is nothing else to a law than a feature of the divine will, the execution of that will requires God to be causally active in every event that 'obeys' those laws.[7]

To think otherwise—that is, to assume that Descartes's God could fix the laws of nature in such a way that they could operate on their own, as it were, without a supervenience base in creatures and without God's constant activity—is flatly inconsistent with the very nature of those laws. How, then, can we explain the

---

[7] Something like this argument might be implicit in Garber (1992).

tendency of commentators to miss this point? One possibility is that philosophers are importing into the discussion a roughly contemporary position on the laws of nature, according to which these laws, once fixed, can (metaphorically) govern on their own. For my part, I find such a conception obscure; what is more important, Descartes simply does not have it.

Helen Hattab offers an intriguing version of the concurrentist reading, which might be capable of parrying this argument.[8] On her view, as on mine, Descartes's secondary causes are the laws of nature. Though she calls this a 'controversial reading,' it is simply what Descartes asserts in the *Principles*. Unlike me, however, Hattab reads this as a statement of what she calls 'Jesuit scholasticism,' as found in such writers as Suárez. This is puzzling, since, as I have been at pains to make clear, there is a vast difference between secondary causes as laws and as powers or modes of created beings. If secondary causes were literally laws, there would be no sense to be made of God concurring with these laws in a way even roughly analogous to that in which he concurs with the powers of bodies on the scholastic view. But as Hattab reads Descartes, 'The laws effectively function like a universal form of *res extensa*.'[9] This is not right, since the laws simply have nothing to do with the nature of extended substance. That is, they 'prescribe' its behavior, but they in no way follow from its essence. And even if this point were brushed aside, it remains the case that the laws are not functional replacements for forms, except in the trivial sense that they represent a vastly different way of conceiving how the natural world works.

My argument from Descartes's conception of the laws of nature to body–body occasionalism is hardly original. In 1678 Ralph Cudworth mounts an argument that has much the same upshot. And coming as he does from the same time period and roughly the same intellectual context, Cudworth's text provides a helpful window through which to see Descartes's own work.

Although copious to the point of monomania in both naming and explicating the views of other authors, Cudworth does not in this section of *The True Intellectual System of the Universe* refer explicitly to Descartes. There can be little doubt, however, that the Cartesian system was at least among those in his mind when he writes of 'those mechanic Theists'

who rejecting a plastic nature affect to concern the Deity as little as is possible in mundane affairs, either for fear of debasing him, and bringing him down to too mean offices, or else of subjecting him to solicitous encumberment; and for that cause would have God to contribute nothing more to the mundane system and economy, than only the first impressing of a certain quantity of motion upon the matter, and the after conserving it, according to some general laws; these men, I say, seem not very well to understand

---

[8] See Hattab (2001, 2007).

[9] Hattab (2007: 64). A similar view is offered by Flage and Bonnen (1997: 842), who argue that 'Cartesian natural laws are ontologically and epistemically indistinguishable from eternal truths: they constitute the form of the world.'

themselves in this. Forasmuch as they must of necessity, either suppose these their laws of motion execute themselves, or else be forced perpetually to concern the Deity in the immediate motion of every atom of matter throughout the universe, in order to the execution and observation of them. The former of which being a thing plainly absurd and ridiculous, and the latter, that which these philosophers themselves are extremely abhorrent from, we cannot make any other conclusion than this, that they do but unskilfully and unawares establish that very thing, which in words they oppose [i.e., the hypothesis of a plastic nature].[10]

Cudworth presents a trilemma for the 'mechanic Theist.' Such a philosopher must either (a) be an occasionalist, (b) suppose that the laws of nature can 'execute themselves,' i.e., function independently of God, or (c) cry uncle and accept Cudworth's own plastic nature.

Option (b) is 'plainly absurd and ridiculous.' But why? Only if Cudworth is here thinking of laws of nature on the divine command theory model, as directives. No command can, by itself, bring about its own execution; it requires a further agent to put it into effect. The only way to resist this would be to argue that it is the objects themselves who put these commands into action. But, Cudworth argues, the material world on the mechanist picture is barren of intellect and will and so cannot be expected to behave itself in accordance with natural laws. In a passage that might have been written by Boyle, Cudworth argues that, although 'the works of nature are dispensed by a Divine law and command, yet this is not to be understood in a vulgar sense, as if they were all affected by the mere force of a verbal law or outward command, because inanimate things are not commandable nor governable by such a law.'[11]

Cudworth concludes that mechanists must (c) endorse his own plastic nature. '[T]he laws and commands of the Deity . . . ought not to be looked upon, neither as verbal things, nor as mere will and cogitation in the mind of God, but as an energetical and effectual principle, constituted by the Deity, for the bringing of things decreed to pass.'[12] But as (c) was not available to Descartes (and in any event would have been rejected by him as a return to the scholastics' 'little souls'), and (b) is out of the running, only (a) is a real alternative as an interpretation. On all three interpretive questions I have posed above, concurrentism earns a 'no': it is not a view the mature Descartes is committed to, whether explicitly or implicitly, nor are any features of it entailed by his mature view.

9.3 THOROUGHGOING OCCASIONALISM

In so far as the natural world is subject to laws, its behavior will have to be caused by God alone. This position leaves open whether finite minds can be causes, which is just as well, since Descartes clearly holds that they can.

---

[10] Cudworth (1837: i. 213–14).   [11] Cudworth (1837: i. 209–10).
[12] Cudworth (1837: i. 228–9).

There is another argument in the wings, however, one that yields not limited but thoroughgoing occasionalism. As I shall examine it in much greater detail below, in connection with Malebranche, a brief statement here will have to suffice. (After all, it is not to be found in Descartes's own works, except perhaps implicitly.) As we have seen, Descartes's God is responsible for the *esse* of all created beings. As Descartes puts it in the Third Meditation: 'it does not follow from the fact that I existed a little while ago that I must exist now, unless there is some cause which as it were creates me afresh at this moment—that is, which preserves me' (AT vii. 49/CSM ii. 33). Although Descartes does not say so here, the point holds for all created beings. What has come to be known as the 'divine concursus argument' takes this as its starting point. For if God must re-create the world at each moment, must he not create individual objects with fully determinate positions in space, and with a full complement of determinate modes? If so, there seems to be no way for any created being, whether mind or body, to contribute to the course of natural events.

A better way to frame this debate is to begin by asking, not how God re-creates the world at each moment, but how he serves as a cause of motion.[13] On the view suggested by the divine concursus argument, which has come to be known as the 'cinematic view,' God's causing motion is just his re-creating objects in new locations. If the cinematic view were right, each moment would be like the frame of a film, disconnected from any other frame except by its place in the sequence. What is more, God would have taken over the role of created minds in causing bodies to move, since as the director (or better, stop-motion animator), he would have to choose where and when to create the bodies in question. Whatever minds wanted to do would be irrelevant, except in so far as God chose to take it into account, making sure that he kept the books balanced where the overall quantity of motion was concerned. While it has its drawbacks as a reading of Descartes—particularly where finite minds are concerned—it is not without its virtues. For example, the cinematic view makes sense of Descartes's rejection of real qualities. Motion, *qua* mode, cannot literally be transferred from one body to another. So if we are to take Descartes's ontology of motion at face value, we need some alternative story to explain the apparent impacts and collisions of bodies besides the migration of modes. The cinematic view does this nicely. For in that case, no motion is changing hands; talk of transference is just a handy way of talking about God's re-creating objects in successive locations.[14] This is a substantial, and perhaps decisive, consideration in favor of the cinematic view.

I think the divine concursus argument in fact raises two very different issues I have so far lumped together. On one hand, the argument in the hands of

[13] One advantage of my strategy is that it allows us to ignore a possible complication in the divine concursus argument, namely, the fact that Descartes only says in Meditation III that God must 'as it were' re-create me. Although I am not inclined to make much of this phrase, some might be.

[14] The problem of what to make of the transference of motion (or 'impulse') will be with us throughout; see esp. Ch. 15.

Malebranche and Louis de La Forge concerns the *locations* of bodies. These are extrinsic properties of bodies. The argument so formulated thus says nothing directly about God's responsibility in creating the modes of beings. For modes are intrinsic properties. A different issue is raised by the necessity for God to re-create each substance at each moment with its full roster of modes. This is a distinct point: whether or not God must choose the locations of bodies, he must also, it seems, choose which modes to preserve and which not. On the second version of the argument, God is straightforwardly responsible for *all* the modes of bodies, while the first line of thought suggests that God is responsible for their locations, with the result that there is nothing for bodies to contribute by way of producing or distributing motion.

Let us consider the second version: can Descartes's God re-create substances without at the same time creating their modes? The answer is no, and Descartes's ontology provides the reason. Now, on the cinematic view, God is responsible for not just the being but all the modes of bodies, plus their locations. Note that it is agreed on all hands that Descartes's God must be responsible for the being of created substances. To resist the cinematic view, one would have to come up with a way for God to preserve only the being of a created substance while leaving (at least some of) its modes up to the activity of other such substances. But this requires reifying modes in a way inconsistent with Descartes's ontology. It is not just that Descartes rejects real qualities, which can exist independently of the substances they modify. His conception of the substance–mode relation, as developed above, precludes the bifurcated contribution of God and creatures to being and mode, respectively. For at any one moment, there are not *two* things, the substance and its modes; there is just the determinable existing in various determinate ways, and motion is one of those ways or modes. And unless these two could be somehow prised apart, there is no way to make sense of God being responsible for one, and bodies, the other.[15] So even if the cinematic view conflicts with mind–body causation, it is hardly without textual support.

In either form, the divine concursus argument threatens to make all created beings causally impotent. The proponents of both concurrentism and limited occasionalism have attempted to parry these lines of argument by distinguishing between God's role as cause of motion and as preserver of finite beings. Garber, for example, claims that God acts as the cause of motion by supplying 'a divine shove.'[16] If this is right, then it leaves open the possibility that created minds act, by similarly supplying a 'mortal' shove. The key lies in allowing that God's conservation of motion applies only to the motion that he himself introduced at the beginning of the world; created minds (and perhaps angels) are free to

---

[15] This makes trouble for Clatterbaugh's suggestion that Descartes avoids the divine concursus argument by making God the preserver of substances but not their modes. See Clatterbaugh (1999: 118 ff.). We shall return to this issue below, sect. 10.3.   [16] Garber (2001: 201).

add what they will.[17] What God does, in virtue of his immutability, is simply conserve the initial quantity of motion with which he began at the first moment of creation. Whether other beings add motion or not is neither here nor there. Thus, there is no conflict between minds introducing motion to the world and the conservation law, since what is conserved is not motion *tout court* but the motion God himself causes.[18] For this to make sense, God must conserve some bodies at some times without causing them to move, allowing created minds to do so instead.

The core question, then, for this attempt to defuse the divine concursus argument is whether Descartes does wish to distinguish between God's roles as continuous creator of bodies and as cause of motion in the way the 'divine shove' reading requires. I don't think so.

Consider how Descartes deduces the conservation principle in the section immediately preceding the discussion of laws of nature. Descartes infers from divine immutability that 'God imparted various motions to the parts of matter when he first created them, and he now preserves *all this matter in the same way, and by the same process by which he originally created it*; and it follows from what we have said that this fact alone makes it most reasonable to think that God likewise always preserves the same quantity of motion in matter' (AT viiiA. 62/CSM i. 240; my emphasis). God's act of preserving the matter he created is precisely the same as his act of conserving the same quantity of motion in it. For it is only because God preserves matter in the same way as he did at the first moment of creation that we can infer that it contains the same quantity of motion. The constant quantity of motion in the world is achieved, not by God's continuing to shove matter around, but simply by the *way* in which he preserves matter.

Descartes elaborates on this point in his derivation of the second part of Law 3:

the whole of space is filled with bodies, and the motion of every single body is rectilinear in tendency; hence it is clear that when he created the world in the beginning God did not only impart various motions to different parts of the world, but also produced all the reciprocal impulses and transfers of motion between the parts. Thus, *since God preserves the world by the selfsame action and in accordance with the selfsame laws as when he created it*, the motion which he preserves is not something permanently fixed in given pieces of matter, but something which is mutually transferred when collisions occur. (AT viiiA. 66/CSM i. 243; my emphasis)

God not only preserves the world by re-creating it at each moment; he does so according to the laws. But what is it to re-create a bit of matter in a new place

---

[17] See Garber (1992, ch. 9).

[18] Other commentators have taken different paths toward reconciling the introduction of motion by minds with the conservation law. Leibniz, for example, suggests that, on Descartes's view, minds can alter the direction, but not speed, of motion. If so, minds would not, it seems, violate the conservation law when they intervened in the world. But as commentators have pointed out, there is insufficient textual support for this claim. See esp. Remnant (1979) and Schmaltz (1996: 213–15).

according to the laws of motion, and then add a 'divine shove'? When God first created the world, he created it with bits of matter in motion. And this text clearly states that God *goes on* doing exactly the same thing. So far from divorcing God's role as cause of motion from his role as preserver of the universe, Descartes seems to wed them so tightly they become identical. God, '*by the same action* by which he keeps matter in existence . . . also preserves as much movement' as he at first created (Letter to the Marquess of Newcastle, AT iv. 328/CSMK 275; my emphasis). There are not two activities here—preserving matter in existence and preserving its motion—there is only one. Thus, there cannot be a distinction between God's activity as source of motion and as preserver of beings. Just as the divine concursus argument requires, these activities are one and the same.

This textual evidence, however, has failed to persuade some of the most acute recent commentators, including Garber and Pessin.[19] And the basic point, at least for Pessin, is quite simple: since God need not have created matter-in-motion but could have created stationary matter, the two acts—creating beings and causing motion—must be distinct. Pessin quotes Des Chene's argument that since 'the existence of motion [is] not entailed by the existence of matter, as opposed to the possibility of motion, the creation of motion is independent of the creation of matter.'[20] Thus, God's conservation of motion is a distinct act from his conservation of bodies.

Descartes clearly holds that motion must be added to the world by God, even given his creation of matter, since motion does not follow from the nature of matter. Thus, the creation of matter can take place independently of that of motion, though not, of course, the other way round. And this simple fact, it seems, should be enough to show that the two creative acts are not identical, despite Descartes's apparent claims above.

I find this argument unpersuasive. While it is perfectly true that creating matter does not entail creating motion, this does not show that God does not, by a single volition and action, accomplish both. Moving chess pieces about on the board does not entail that I am playing chess; but that does not mean that playing chess is a further activity over and above moving the chess pieces. In fact, this example exhibits precisely the modal asymmetry of God's creation of motion: just as I cannot play chess without moving the pieces, so God cannot create motion without creating matter. He can, however, create matter without creating motion, just as I can move the pieces about without playing chess. So the two act-types are not identical. But of course their tokens can be. All of this tells in favor of the cinematic, as opposed to the divine shove, model of divine causation.[21]

---

[19] See Garber (1992: 275 ff.) and Pessin (2003: 27). Des Chene might be reckoned among their number; Pessin cites him as his source for the simple argument I present below.

[20] Pessin (2003: 27); the quotation is from Des Chene (1996: 329), though I am not sure Des Chene would agree with the use Pessin makes of it.

[21] I should point out that I reject any connection between the doctrine of continual creation and temporal atomism: the two issues simply have nothing to do with one another. However thick the

But perhaps there is other internal evidence against the cinematic view. Helen Hattab argues that this view is inconsistent with divine immutability.[22] As Garber puts it, 'if motion is nothing but God's sustaining bodies in different places at different times, then, it would seem, a world sustained by an immutable God should be a world without motion.' Hattab takes this to be the strongest reason for rejecting the cinematic interpretation. But Garber himself goes on to argue that, on whatever reading one adopts, Descartes cannot make the existence of motion consistent with that of an immutable God. 'Descartes simply fails to reconcile the transfer of motion from one body to another with the immutability of God. In the end, it remains a puzzle why an immutable God should create a world in which, from the beginning, change is such an integral part.'[23] A point that tells against any of the possible interpretations cannot decide among them.

At this stage of the debate, there is a further argument my opponents can muster. In the Second Replies, Descartes writes that it is 'greater to create or conserve a substance, than it is to create or conserve the attributes or properties of a substance' (AT vii. 166/CSM ii. 117).[24] This certainly implies some sort of distinction between creation of a being and creation of its modes. This passage seems to allow that creatures could 'create or conserve' a body's modes, while God does the same with regard to the substance *tout court*.

My present argument, however, does not turn on collapsing the distinction between creating a substance and creating its modes. The distinction I have in my sights is that between creating *motion* and creating substances. So in this context, the Second Replies passage is something of a red herring. Even if it weren't, however, any argument built on it would be lame. For it would be vulnerable to precisely the objection I lodged above: to say that the one act is greater than the other is not to say that they must be distinct. Playing (or winning at) chess is 'greater' than moving my pieces around the board. But that doesn't imply that in playing (or winning) I am doing anything more than moving my chess pieces.[25] We must keep in mind the context in which the Second Replies passage occurs. The axiom it embodies is preceded by Axiom VIII, which states that 'whatever can bring about a greater or more difficult thing can also bring about a lesser thing.' If anything, then, the lesson to be drawn from these two axioms favors the occasionalist reading, for it is an argument for the claim that God can create not only a thing but its modes as well.

---

time-slices during which God re-creates the world turn out to be, whether they are appropriately called 'instants' or 'moments,' is neither here nor there. So I do not think that the view I have sketched is committed one way or the other on that issue. (I owe this point to Jorge Secada.)

[22] Hattab (2007: 60); she quotes Garber (1992: 282).

[23] Garber (1992: 291).   [24] See Garber (1992: 277).

[25] Of course, I have to move them *in a certain way* to win. But this is precisely my point: I accomplish the one act by means of the other, just as God brings about the motion of objects by means of his continual re-creation of them.

If the distinction between the volitions (and acts) by which God continually re-creates matter on one hand and those by which he creates motion on the other goes by the board, the linchpin of the argument against the cinematic view is removed. And once this happens, attributing a thoroughgoing occasionalism to Descartes becomes inevitable.

9.4 THE PROBLEM OF MENTAL CAUSATION

If I am right, there is a core tension in Descartes's mature position, one that resists all attempts to read it away. With regard to body–body causation, Descartes is an occasionalist. He cannot be called a thoroughgoing occasionalist, however, since his belief in the causal power of minds, not just God's but our own, is non-negotiable. Writing to More in August of 1649, Descartes says that the power causing motion may be 'God himself preserving the same amount of transfer in matter as he put in it in the first moment of creation; or it may be the power of a created substance, like our mind' (AT v. 403–4/CSMK 381).[26]

It is important to see that the problem of mental causation applies no less to God than to finite minds. The clearest way through the thicket would seem to be to begin with God's activities and work our way back to our own. I have been suggesting that Descartes's continual creation doctrine commits him to the cinematic model. The most influential alternative reading, Garber's 'divine shove' interpretation, takes its inspiration from an earlier letter of 1649, where Descartes offers to explain God's role as a cause of motion in terms of our own. Note that this feature is conspicuously absent from the August letter, also to More; in the passage just quoted, for example, Descartes is at pains to show the wide gulf between the powers of created substances and that of God, in so far as God's role as cause of motion reduces to his preservation of the same 'amount of transfer,' which is not something we finite minds are capable of. The transfer—i.e., the motion—is a mode of a body, and to preserve the same amount from eternity is to create those modes in a uniform way.

In this earlier letter to More (April 15, 1649), Descartes says, 'I must confess that the only idea I can find in my mind to represent the way in which God or an angel can move matter is the one which shows me the way in which I am conscious I can move my own body by my own thought' (AT v. 347/CSMK 375).[27] This passage is the inspiration for Garber's 'divine shove' reading. But in this same passage, Descartes denies that 'any mode of action belongs univocally to both God and his creatures.' So my experience of willing my arm to move is

---

[26] The text continues, 'or of any other such thing to which he gave the power to move a body.' While this might be either an admission of the power of angels to move bodies or mere deference to God's omnipotence, it does not carry the implication that bodies can have this power.

[27] Garber also uses this passage in several contexts.

not a model for God's activity, except in some unspecified analogical sense. All of this makes it obscure just how my conceiving of my own mind's activities helps me to understand anything at all about God's. Descartes might be presupposing a proportional theory of analogy *à la* Aquinas and Cajetan, but, as Geach has shown, this kind of analogy is unable to provide meaning to claims about God.[28] That is, even if Descartes does indeed mean to invoke proportional analogy, it is in fact no real help in understanding how God acts.

Surely Descartes does not mean to tell More that human minds act on the world by re-creating it at each moment, or by conserving the modes of bodies. But what then could he mean? Here it is important to keep in mind that Descartes does not say that the idea he forms of his own mind's action represents (analogically or otherwise) the way in which God acts; he makes instead the much weaker claim that this is 'the only idea [he] can find in my mind to represent' God's activity. *Faute de mieux*, this idea will have to serve.[29] I suspect that Descartes simply has nothing informative to say on the matter. If so, the letter to More is the closest he comes to a confession.

Thus, the comparison between God and finite minds does not help much, at least as an answer to our question how God causes motion. First, it assumes that we have a firm grasp of what is going on in our minds when we exercise our causal power. Second, it plays down Descartes's own hesitancy about likening God's activities to our own. If there really is no univocal sense of 'action' that can be applied both to minds and to God, how are we to understand what God is getting up to when he acts on the world? We must keep in mind that this is a letter and not a fully worked out statement of doctrine. If Descartes himself began to feel qualms about the utility of the comparison, this would explain why there is no sign of it in his letter five months later to the same correspondent.

The further problem, of course, is that Descartes does not do much to help us understand how our own minds act on the world.[30] On my view, it is not surprising that Descartes has little to say to explicate the power of finite minds. The real problem with understanding how a Cartesian mind causes anything is not that minds and bodies are of two very different categories—minds are not extended in space, while bodies are, for example—but rather that a Cartesian mind is an Aristotelian substantial form, and is no more or less intelligible in Descartes's context than any other. Just like the Aristotelian's heaviness, a Cartesian mind can somehow suffuse or be fully present in a body (and so it is in this sense 'in space'). A mind can, like heaviness, be directed at non-actual

[28] See Geach (1980).
[29] Cf. Leibniz, who, discussing Descartes's position on mind–body interaction, writes, 'Descartes had given up the game at this point, as far as we can determine from his writings' (AG 143).
[30] There is the pineal gland as a possible explanans, but this of course just pushes the problem back. There is also the suggestion that a Cartesian mind might change the direction, but not the amount, of motion in the pineal gland, thus preserving the conservation law. But as Remnant (1979) persuasively argues, this is not to be found in Descartes's texts.

possibilia. And of course the scholastics' heaviness and minds alike can change the course of events, at least in the bodies they enform. This gives another reason why understanding God's activity on analogy with our own is bound to be a non-starter—unlike God's mind, ours (at least at the moment) enforms a body. The only real alternative here would be to abandon the notion of the 'divine shove' altogether and rest content with God's continual re-creation. And this is precisely what Malebranche will do.

PART II

THE DIALECTIC
OF OCCASIONALISM

INTRODUCTION

Ontological mechanism begins with the impulse to clean up the Aristotelian's excesses. So far, this has led to the rejection of course-of-nature mechanism. For the key notion that underwrites the Aristotelian bottom-up picture, that of power, seemingly has no place in a world of reified geometrical objects. What is more, Descartes's invention of the laws of nature also entails a rejection of the bottom-up model, since those laws require God's ubiquitous activity for their enforcement. At the same time, one of the driving forces behind ontological mechanism was the desire for perspicuous causal explanations, and in particular, causal explanations that would appeal to the newly stripped-down natures of created beings.

None of this is to say that Descartes represents a complete departure from the Aristotelians. Indeed, it is precisely because he preserves core elements of the Aristotelian view, such as the *esse-ad* requirement, that he finds it impossible to reconcile ontological and course-of-nature mechanism. Nevertheless, occasionalism is at best implicit in Descartes's works. Thus, we begin this part of the book by exploring the arguments of the most prominent modern defender of thoroughgoing occasionalism, Nicolas Malebranche. Refining Descartes's own 'little souls' argument, Malebranche develops what I call the 'cognitive model' of causation. The antecedents of this model can be traced further, back to the Aristotelian's own *esse-ad* requirement. The directedness of genuine causal power, Malebranche thinks, requires the kind of connection that only intentionality can supply.

In the final two chapters of Part II, I turn to a quite different response to the Cartesian predicament. While Malebranche takes the inevitable result of ontological mechanism to be a strong top-down picture of causation, Pierre-Sylvain Régis fights to preserve the causal powers of bodies. To do this, he must forge a new model of causation, the geometrical model. In England, Locke and others had already done this in their own ways, decades before Régis publishes his work. But working within the Cartesian tradition, and at the same time self-consciously trying to preserve scholastic concurrentism, Régis represents the next logical move in the debate.

Before any of this can be brought into clear focus, we must look at Malebranche's arguments. For it is only by replying to them that the proponents of the bottom-up picture can hope to preserve the causal powers of bodies and forge their own model of causation.

10

Malebranche and the Cognitive Model of Causation

Although his arguments have been widely discussed, it too often remains obscure why Malebranche would think they could be at all persuasive. By the end of this chapter, we shall see that the negative thesis of occasionalism—finite beings cannot be causes—is an all but inevitable result of combining key elements of a scholastic conception of causation with the mechanist ontology.

I shall argue that Malebranche's dismissal of bodies as causes makes sense only if the requisite tie between cause and effect involves intentionality. Only by means of this kind of intrinsic directedness can an object or event pick out or be directed toward its cause. Having taken over key elements of the scholastic conception of causation, Malebranche finds that in the context of mechanism nothing but God's will can fit this conception. Several of Malebranche's equally obscure arguments, particularly those that deny causal powers to finite minds, bear the stamp of this line of thought.[1] Malebranche (following Descartes) rejects the attribution of powers to bodies on the grounds that *esse-ad* amounts to intentionality, a feature only minds possess. The flip side of this, however, is that causes and effects need precisely the kind of tie intentionality alone can provide. What makes a divine volition a suitable causal relatum is the intentional nature that ties it to its effects, since the propositional content of a divine volition *just is* that volition's effect. This cognitive model of causation is the foundation of the 'no necessary connection' argument (or NNC).

The cognitive model can help us understand another of Malebranche's puzzling arguments. Finite minds, he claims, cannot cause physical events. Although it is clear that any finite mind, lacking omnipotence, cannot live up to the demand that a cause be necessarily connected with its effect, Malebranche does not rely solely on this consideration. Instead, he offers what I shall call the 'epistemic

---

[1] I do not mean to suggest that the line of thought I develop below, which links at least three of Malebranche's arguments (what we shall call the 'epistemic argument,' the 'little souls argument,' and NNC), lies behind, or is directly connected with, other argumentative strategies one can discern in the texts. I am of course aware, for example, that Malebranche offers several explicitly theological arguments, including the argument from leeks and onions: Malebranche argues that attributing true causal powers to bodies would require us to worship them, and thus commits us to paganism. I shall not discuss such arguments here.

argument': if a mind were to cause the motion of, say, one's arm, it would have to will the temporal antecedents of that event, which include brain events. But we seem to move our arms all the time in the absence of such knowledge; so whatever the cause is, it cannot be our minds.

The usual story about this argument is that it relies on Malebranche's doctrine of blind will. For Malebranche, the will is a blind faculty that requires the understanding to direct it. Thus, human minds cannot cause physical events for want of the knowledge of the workings of the animal spirits and nerves. But, as I shall show, the doctrine of blind will goes no distance at all toward justifying Malebranche's conclusion. If the mind cannot will the motion of the animal spirits as such, it can surely will the motion of one's limbs. Why should we believe that a mind must will all the antecedents of a physical event as such in order to effectively will the result? In section 10.6, I show how the epistemic argument comes into focus only if we assume, with Malebranche, that a cause must include its effect as its intentional object. This alone lets us see why Malebranche thinks finite minds cannot be causes.

10.1 THE ARGUMENT FROM NONSENSE

The opening salvo of Malebranche's attack on secondary causes in Elucidation XV is what Irving Thalberg has called 'the argument from nonsense.'[2] Such an argument seeks to defeat a claim not by directly attacking its justification but by exposing that claim as gibberish. Malebranche goes so far as to say that it is 'the principal' reason 'preventing [him] from attributing to *secondary* or *natural* causes a force, a power, an efficacy to produce anything' (*SAT*E XV: 658).[3]

> this opinion does not even seem conceivable to me. Whatever effort I make in order to understand it, I cannot find in me any idea representing to me what might be the force or the power that they attribute to creatures. And I do not even think it a temerarious judgment to assert that those who maintain that creatures have a force or power in themselves advance what they do not clearly conceive. (*SAT*E XV: 658)

What's wrong with concurrentism is not a mere problem of detail; the whole position is simply nonsense. For if we understood 'power' as applied to bodies, we would have to have an idea of it, or, in Malebranche's terms, God would have to reveal this idea to us. But search as he might, Malebranche cannot discover any such idea. And 'being a man as much as they' (*SAT*E XV: 658), he ought to be able to find it if they can. Moreover, the mere fact of disagreement about the nature and status of secondary causes among the scholastics is indirect evidence

---

[2] See Thalberg (1981).
[3] Here Malebranche echoes Descartes's argument from obscurity; see AT iii. 649/CSMK 216.

that they have been talking nonsense all along.[4] As Hume was later to argue in the case of free will, the persistence of a seemingly unresolvable debate is an indication that the participants are either talking past each other or not *talking* at all but merely producing sounds.[5]

Note that, despite appearances, the argument does not begin with the claim that power is inconceivable and then infer that it is impossible. Although Malebranche, as we shall see, deploys the conceivability principle (if *x* is conceivable, *x* is possible), there is no evidence that he holds the converse, which would be implausibly strong. The claim is instead that attributing powers to bodies is sheer nonsense, sounds without meaning.

But isn't this moving from an epistemic claim to a metaphysical one? Why couldn't the partisan of secondary causes admit the obscurity of the notion and still insist on the reality of the thing? In the seventh of his Dialogues on Metaphysics, Malebranche has Aristes put just this objection to Theodore:

ARISTES. We do not know, it will be said, what that power is. But what can you conclude from an avowal of ignorance?

THEODORE. That it is better to say nothing than not to know what one is saying.[6]

Although this argument becomes one of the chief weapons of the empiricists,[7] it is worth noting that a paradigmatic rationalist is among the first to deploy it. In fact, it is the direct descendant of Descartes's insistence on the way of clear and distinct ideas as the only means to knowledge. Now, Malebranche does not conclude that, as it happens, we do not have an idea of power. It is not a contingent fact at all, from Malebranche's point of view, that we lack this idea. For no such idea is in principle possible. As I shall show, neither bodies nor minds are the right sorts of things to have powers, on Malebranche's view. Thus, the argument from nonsense does not stand on its own; it is best regarded as a compressed anticipatory summary of several other lines of thought Malebranche deploys.

10.2 THE ARGUMENT FROM ELIMINATION

In the seventh of his Dialogues on Metaphysics, Malebranche's spokesman Theodore responds to Aristes's claim that bodies can act on minds:

. . . I do not understand . . . how bodies can acquire in themselves a certain power by the efficacy of which they can act on our minds. What would that power be? Would it be a substance or a modality? If a substance, then it is not bodies that act but that substance which is in bodies. If the power is a modality, then there will be a modality in bodies

---

[4] See *SAT* E XV: 659.   [5] See esp. *E* 8.
[6] Malebranche (1992: 224).   [7] See Ott (2004*b*).

which is neither motion nor shape. Extension will be able to have modalities other than relations of distance.[8]

The defender of secondary causes faces a dilemma: the power in virtue of which the body is a genuine efficient cause must be either a substance or a mode. If that power is a substance, then the body itself cannot be the cause. If it is a mode, it must be something in the body that is neither motion nor shape.

It is difficult to assess the effectiveness of this argument. Let us take each horn separately. Is there anyone among Malebranche's targets who would embrace either of them? Consider the view that powers are real qualities, i.e., qualities that can migrate from one body to another. Although this is not Aristotle's own position, or even a necessary part of medieval concurrentism, it might have had some appeal in the context of the debate over transubstantiation. Briefly, the idea would be that the form of the bread can come and go, depending on the stage of the Eucharistic ritual. If one simply identifies the form of the bread with, say, its power to nourish, then one would be thinking of at least this power as a real quality. One would then be not so far from holding that the bread's power to nourish is itself a substance, alongside the bread. If we plug in the Cartesian argument against real qualities—they are supposed to be forms, but since they need not inhere in any subject, they have characteristics applicable only to substances—we can see how at least by Malebranche's lights his opponents would have transformed a power into a substance.

Whatever one makes of this, it seems clear that Malebranche can argue successfully against this position. After all, powers were initially invoked to explain why *substances* can do what they do. As forms or entailments of forms, they explain why a given body behaves as it does and, what is more important, they explain this by means of features *of* that body. If powers are made into substances in their own right, however, this key feature is lost: it is purely contingent that power-*qua*-substance x is attached to substance y. Thus, embracing this horn of the dilemma sacrifices the very feature that made the Aristotelian bottom-up strategy attractive in the first place.

When we turn to the second horn, matters become more complicated. Later in Part II, I will show how a mechanist concurrentism such as that of Régis, which takes this second horn as a way out of the dilemma, has the resources to answer Malebranche's argument. But let us consider Malebranche's point on its own. The problem he sees with the second horn is that it would introduce a new mode alongside shape or motion. It is not clear why this should be so. Why think that a body's motion and shape should not be sufficient to explain its acting on another substance? That is, why not try to reduce power to the categorical and intrinsic modes of bodies?

---

[8] Malebranche (1992: 224). Although Theodore raises this question with regard to body–mind causation, the argument tells equally against body–body causation. The same argument occurs at *SAT* I. x. 1: 49.

To see why Malebranche regards this as a non-starter, we need to understand his peculiar conception of the modes of extension. For Malebranche, all the modes of extension, motion and shape included, are reducible to relations, in particular, relations of distance. Although it remains true that a body's modes just are its *manières d'existence*, just as they are for Descartes, these different ways of being are identified by Malebranche with relations of distance. As far as I can see, there is no real argument for this; in the *Dialogues*, for example, Theodore asks Aristes to 'consider attentively the clear idea of extension,' and then asks him whether it is not evident 'that the only possible properties of extension are relations of distance.'[9] Some of these relations are fixed, e.g., size and shape, whereas motion is successive. I suppose the point here is that for x to move from a to b in the interval $t-t'$ is for x to have a different distance from a and b at t' than at t, whereas for x to be shaped like a square is only for certain points along its perimeter to be related in a given way. Thus, motion, unlike shape, includes a temporal element, and this is what it means to say that motion is 'successive and always changing.'

In the case of the stable modes, Malebranche's reduction to relations of distance is much less plausible. Consider again the analysis of squareness. We are invited to think of a set of points along the boundary of the body; that body is square just in case those points bear certain relations to each other. This is fine. But how are we to understand those points themselves? How can they be reduced to relations of distance? Without an objective grid, as it were, on which to plot them, it seems to become a matter of decision just where the boundaries of an object, and hence the points selected for a determination of shape, lie. Now, this grid might be provided either by attributing to the points some kind of intrinsic, non-relational property, or by the relation of part of the body to space itself. But the former will clearly not do. What kind of non-relational property would turn the trick? And the second is blocked to Malebranche, since, like Descartes, he identifies space with extension. Any proposal along these lines will then turn out to be circular.

More fundamentally, we can see that something has gone horribly wrong in the ontology of extended substances and their modes if modes get reduced to relations. For any relation, it makes sense to ask what its relata are. If these turn out to be relational themselves, we are embarked on a regress. Without some ground in a non-relational property, the modes of extension are mere structure without stuffing.[10]

If, however, we grant Malebranche his conception of the modes of extension, his argument goes through. For it is indeed hard to see how a power could consist in nothing more than a set of relations of distance. But his reliance

---

[9] Malebranche (1992: 173).

[10] It has been suggested to me that Malebranche might identify 'relations of distance' with 'relations of distance between any material parts.' But this just pushes the problem back: what are we to make of these material parts themselves?

on this conception in the elimination argument clearly weakens his position dialectically. For the relational account of modes is idiosyncratic, peculiar not just to Cartesianism per se but to Malebranche's own version of Cartesianism. To this extent, the argument loses its force against opponents who do not share it. Let us turn then to another occasionalist argument.

10.3 THE DIVINE CONCURSUS ARGUMENT

Malebranche argues in Dialogue VII that 'it is a contradiction—a "contradiction," I say—that bodies can act on bodies.' Here is his summation:

Creation does not pass: the conservation of creatures is on the part of God simply a continued creation, simply the same volition which subsists and operates unceasingly. Now, God cannot conceive, nor consequently will, that a body be nowhere or that it not have certain relations of distance with other bodies. Hence, God cannot will that this chair exist and, by this volition, create or conserve it without His placing it here or there or elsewhere. Hence, it is a contradiction that one body should move another.[11]

Here is one way to reconstruct the argument:

1. One body can move another, i.e., cause it to occupy a given place. (Assumption)
2. If one body can move another, the volition by which God conserves it in existence must not include any reference to any particular place, on pain of overdetermination.
3. Conceivability entails possibility.
4. Necessarily, any body that exists, exists in some fully determinate place or other.
5. It is inconceivable that a body exist in no determinate place. (4, contrapositive of 3)
6. What cannot be conceived cannot be willed.
7. God cannot will to create a body without willing it to exist in a fully determinate place. (5, 6)
8. God conserves all beings at all times.
9. To conserve a being is to re-create that being.
10. God cannot conserve a being without willing it to exist in a fully determinate place. (7, 9)
11. No body can move another (10, 2)

    Contradiction (1, 11)

---

[11] Malebranche (1992: 230).

Malebranche's point is not that the determinateness of reality directly constrains what God can accomplish. Rather, this determinateness constrains what God can *conceive* (via 3), and hence will (via 6).[12] The challenge it presents, then, is to show not merely how God and creatures can collaborate to produce their effects, but also how God can will to conserve, that is, re-create, each object, without willing it to exist in this or that place. For once he wills their locations, each object will exist in its place, and there will be nothing for bodies to contribute to the proceedings. Note that Malebranche does *not* say that God must continually create a substance and all of its modes; rather, the point is that, for everything God creates, he must will to create it in some determinate place or other.[13]

Although Malebranche's conclusion here concerns only the power of bodies to move one another, the same seems to apply to any non-God cause: it will be just as superfluous, since all the work of moving bodies about takes place at the level of God's continuous creation of substances.

Let us work through the argument. It is helpful to keep in mind the ground common to nearly all the disputants, Cartesian and scholastic alike: God must, at a minimum, conserve the world and all of its substances at every moment of their existence in an act that is only conceptually distinct from his initial act of creation.[14] The divine concursus argument then runs a *reductio ad absurdum* against the premise that bodies can cause other bodies to move.

To suppose that body *a* can move body *b* is to suppose that *b*'s future location is due to the activity of *a*, while, in the context of the continuous creation doctrine, *b*'s existence *simpliciter* is due to God's activity. Now, how precisely are we to characterize God's activity? Whenever God acts, he must form a volition: an act of will directed at a given state of affairs. For convenience, we can call the propositional element of God's volition, as distinct from the contribution made by his will, a p-volition. The argument's key question is this: what would God's p-volitions have to look like, if objects could move one another?

Premise 2 claims that a p-volition could not include reference to any particular place if objects are to be genuine causes. Suppose that God wills (Volition 1) to conserve object *x* in place *a* at time *t*, and (V2) in *b* at *t'*. (It is important that the temporal element be included in the content of the p-volition, since Malebranche's God does not act in time. There is no succession in the divine will, although there can be in the propositional contents of those acts of will.) Now suppose further that object *y* moves *x* to *b* at *t'*. If God's

[12] This is one difference between Malebranche's version of the argument and that of La Forge, which Garber (1992) quotes.

[13] See above, sect. 9.3, for a discussion of these two lines of thought.

[14] Although the claim that there is only a conceptual distinction between creation and conservation is widespread throughout the scholastic period and, of course, is familiar from Descartes's Third Meditation, we shall see that Régis challenges it below, in an attempt to render the divine concursus argument toothless.

conservational p-volitions were fully specified along the lines suggested by V1 and V2, we would have a case of overdetermination. For God would then will that *x* exist in *b* at *t'*, and, as all parties to the debate agree, God's will is necessarily efficacious. In Elucidation XV, Malebranche writes, 'Since God's volitions are efficacious by themselves, it is enough that He should will in order to produce, and it is useless to multiply beings without necessity' (*SAT* 679). The claim is not that parsimony alone entails that only God is a cause.[15] Rather, Malebranche's point is that any causal contribution by creatures will be superfluous *if God's p-volitions are fully determinate*. A p-volition, if bodies are to contribute anything, must be indeterminate; it must be something like, 'that body *x* exists at *t*.'

Premise 3—that conceivability entails possibility—is something that all parties to the debate, including Descartes, hold. Descartes's Sixth Meditation argument for the real distinction between mind and body turns on just this principle. It is so prevalent in the early modern period that in 1739 Hume is able to declare that it is 'an establish'd maxim in metaphysics, *That whatever the mind clearly conceives includes the idea of possible existence*' (*T* 1. 2. 2. 8).

Premise 4 is the intuitive principle that all bodies must exist in some determinate place or other. While we clearly do not know all of their locations, that bodies have them should be uncontroversial. It would be a modal fallacy to infer from this principle that each body necessarily is in the place that it is, of course. The claim is simply that physical reality is fully determinate in respect of location.

At premise 5, things begin to get interesting. Take the contrapositive of 3: if *x* is impossible, *x* is inconceivable. Now, if we accept the determinateness principle (premise 4), we get the result that a body that existed in no particular place is not just impossible but inconceivable.

Clearly, one must be able to conceive of a state of affairs before one can include it among the contents of a p-volition. This is all premise 6 states. The rest of the argument is simply drawing the obvious conclusion: God cannot form the p-volitions required if bodies are to be causes (or even co-causes). The picture of bodies being merely conserved by God and making their own causal contributions is incoherent. Of course, as I have noted, few philosophers wished to defend mere conservationism. But if the argument works, it also shows that the concurrentist picture is incoherent, for this picture still attributes causal

---

[15] By contrast, George Berkeley's version of the argument relies on parsimony rather than on the considerations of overdetermination and modality I think are central to Malebranche. In *De Motu*, Berkeley writes, 'Modern thinkers consider motion and rest in bodies as two states of existence in either of which every body, without pressure from external force, would naturally remain passive; whence one might gather that the cause of the existence of bodies is also the cause of their motion and rest. For no other cause of the successive existence of the body in different parts of space should be sought, it would seem, than that cause whence is derived the successive existence of the same body in different parts of time' (*DM* 34).

powers to objects. And these are precisely what is elbowed out by God's activity of conserving bodies.

We are now in a position to make good on an earlier promise and consider how Descartes stands with regard to Malebranche's own divine concursus argument. Given that Descartes seems to hold many of the premises of the divine concursus argument, especially premise 8, it is tempting to read the divine concursus argument into Descartes, and conclude that he is an occasionalist. As Garber has argued, however, there's no reason to think that Descartes ever drew the occasionalist conclusion, even if he were committed to all of the premises.

But does this mean that Descartes can escape its conclusion? A typical strategy here is to appeal to a distinction, not explicitly found in Descartes, between God's conservation of a substance and his conservation of its modes.[16] If God merely re-creates an object *qua* substance and leaves its modes undetermined, changes in those modes can then be attributed to creatures. We have already seen some problems with this move. What is important in this context is that it provides the wrong sort of answer altogether. For the argument concerns not a substance's modes per se but rather its location. One can remain agnostic about the need for God to re-create a substance and its full complement of modes while insisting that each physical substance necessarily exists in some place or other. (This, by the way, provides a way out of a traditional problem in understanding Malebranche's claim that minds can at least partly determine their own future states. A mind's power over itself need not be inconsistent with the divine concursus argument, since that argument does not entail that, for any substance, God must re-create all its modes. The point instead is that for any material substance, God must create it in some location or other.)

In fact, if we are looking for a way out for Descartes, the best bet would be to see whether he is committed to the determinateness principle (premise 4). If we could make sense of a world in which God preserved a given amount of motion in objects *by* creating them and at the same time allowed minds to introduce new motions in these same objects, we could read him as both consistent and consistently committed to the causal powers of mind. This would require that God sustain some objects (those influenced by minds) without sustaining them in some fully determinate place. Our question then becomes, does Descartes think that all of reality is necessarily determinate?

It is true that Descartes calls extension 'indefinite,' which might seem to threaten my claim that for him extension cannot exist apart from its determinate modes. But to say that extension is indefinite is only to say that however great an expanse of space we imagine, 'there are always some indefinitely extended spaces beyond them, which we not only imagine but also perceive to be imaginable in a true fashion, that is, real' (AT viiiA. 52/CSM i. 232). By contrast with God's positive infinity, the indefiniteness of material substance is simply an absence of

---

[16] Clatterbaugh (1999: 119).

limits. This, clearly, is not the sense of 'determinate' that is at issue in the divine concursus argument. Even taking all of these considerations into account, there is good reason to think that Descartes would grant that any extended substance existing outside of the mind must be fully determinate.

Matters are complicated by Descartes's use of motion to individuate bodies, a move that commentators have deplored almost from the start.[17] 'By "one body" or "one piece of matter" I mean whatever is transferred at a given time, even though this may in fact consist of many parts which have different motions relative to each other' (AT viiiA. 53–4/CSM i. 233). Although God created matter and motion at the same time (AT viiiA. 61/CSM i. 36), Descartes does not say that God could not have created matter first and then introduced motion. Before the introduction of motion, then, matter would not be split off into different bodies, each having its own determinate qualities. Would not matter in such circumstances be a really existing determinable, violating premise 4?

Two points are central here. First, even if matter were to lack such qualities, it is far from obvious that this would be matter in the sense of a determinable quality, extension. For imagine a Cartesian mind encountering such an environment. That mind would, I think, *still* have to deploy its innate idea of extension; this is one consequence of the wax argument. Second, and more importantly, the physical bodies Descartes attempts to individuate through motion are not to be identified with material substances.[18] The distinction between these two notions becomes clear if we keep in mind Descartes's claim that 'each and every part [of extension], as delimited by us in our thought, is really distinct from the other parts of the same substance' (AT viiiA. 28/CSM i. 213). Given the infinite divisibility of matter, each part one can conceive is itself a substance because it is capable of existing independently of the other parts surrounding it. Physical bodies, by contrast, are the sorts of macrophysical objects that figure in the laws of nature. So although motion is required to individuate the bodies that Descartes treats in his physics, it is irrelevant to the question at hand, which concerns material substance(s). The evidence, then, indicates that Descartes accepts premise 4. If there is a lesson here, it is that parrying the divine concursus argument in either of its forms requires the machinery of concurrentism, machinery Descartes simply does not have. When we turn to Pierre-Sylvain Régis, we shall see how an ontological mechanist might hang on to the powers of created beings in the face of God's continual creation of all that exists.

10.4 'LITTLE SOULS' REVISITED

The fulcrum of the next set of arguments is what I shall call the 'intentionality requirement': the requirement that cause and effect be connected by the relation

---

[17] See Garber (1992, ch. 6).   [18] See Garber (1992: 176).

of intentionality. To set the stage for this, it pays to look back, however briefly, at Descartes's 'little souls' argument. As we saw above,[19] Malebranche mounts his own version of this argument, claiming that attributions of power to bodies blur the boundary between bodies and minds. If a body had a power to move in a certain direction, it would have to be able to think of that direction, and have some goal in mind by trying to get there. But Malebranche, I believe, adds much to Descartes's own argument. Some of this will emerge only in the context of the 'no necessary connection' argument below. What we can do here is see how Malebranche's continual insistence on the dependence of everything in the universe, and all of its qualities, on God, leads him to extend Descartes's own insights. While Descartes had of course endorsed, against the empiricist and concurrentist orthodoxy, the Augustinian dependence of knowledge and reality on God, Malebranche takes this much further. In Malebranche's work, Augustine's doctrine of 'divine illumination,' according to which all knowledge directly (and not just through the mediation of innate ideas) depends on God, comes to fruition. Malebranche's thoroughgoing occasionalism must be seen as the ontological mirror of this epistemic position.

In the fifth of Malebranche's *Méditations chrétiennes et métaphysiques*, The Word asks,

Can this body move itself? In your idea of matter, do you discover any power [*puissance*]? You don't respond. But suppose this body truly has the power to move itself; in what direction will it go? At what speed? You fall silent again? 'I mean that body possesses enough freedom and knowledge to determine its own movement and its rate of speed: that it is master of itself.' But watch out lest you embarrass yourself. For, supposing that this body were surrounded by an infinity of others, what must it do when it encounters a body whose speed and bulk are unknown to it? It will give to it, you say, a portion of its moving force? . . . But what part? How will it communicate this part or propagate its motion? Do you understand all of this?   (*OC* x. 47–8)[20]

What is at stake here is, in part, the origin of the orderliness of the world. If it were due to objects themselves, they would have to be possessed not just of power but of clear-headed thinking about the course moving bodies should take. This, I think, is the core intuition behind many of Malebranche's remarks; it will be familiar enough from Descartes's own position on the laws of nature. What Malebranche does with the little souls argument is not so much extend it as connect it with Cartesian claims about the course of nature. Nature would not be as orderly as it is were it not *directly* governed by a divine being.

The flip side of this argument is that minds *are* a possible candidate for genuine causes. But where Descartes accords causal power to both finite and divine minds, Malebranche will restrict it to God alone. Recall that Descartes treats created minds as Aristotelian substantial forms, endowed with their own

---

[19] Ch. 5.   [20] My translation.

powers. Malebranche rejects this. But just *why* he rejects it will tell us much about how Malebranche is, and Descartes is not, conceiving of causation.

10.5 THE 'NO NECESSARY CONNECTION' ARGUMENT

In *The Search After Truth*, Malebranche produces one of his most famous arguments for occasionalism, an argument Hume was to lift and take as his own in the *Treatise*. We have already, I think, removed one barrier to understanding the argument by establishing its target: the myth that mainstream Aristotelianism distinguishes between logical and causal necessity (section 3.1). But necessity is, of course, only one element in Malebranche's definition of a cause as a *liaison nécessaire*. Commentators have tended to pass over the other, namely, that there be a *liaison* in the first place. I shall argue that lying behind NNC is a stronger, but admittedly stranger, argument to the effect that any genuine connection between events requires intentionality. If I am right, we shall be able to better understand both NNC and Malebranche's deployment of the 'little souls' argument.

We should begin with Malebranche's quick statement of the argument:

A true cause as I understand it is one such that the mind perceives a necessary connection between it and its effect. Now the mind perceives a necessary connection [*liaison nécessaire*] only between the will of an infinitely perfect being and its effects. Therefore, it is only God who is the true cause and who truly has the power to move bodies.   (*SAT* vi. ii. 3: 450)

For any two finite objects or events *a* and *b*, a causal connection between them could obtain only if those events were necessarily connected. But if there were such a necessary connection, it would be impossible to conceive of *a* occurring without *b* (since conceivability entails possibility, premise 3 from the divine concursus argument). God's will and its effects aside, we can always conceive of this happening; thus, there is no necessary connection, and hence no genuine *causal* connection, between *a* and *b*.

Malebranche's *liaison nécessaire* criterion is, as I have argued above, not an aberration or invention of his own. NNC's real target is no straw man but a widespread scholastic picture according to which a total cause (God's concurrence, plus the relevant active and passive powers) logically necessitates its effect.[21] It is thus an analytic truth that fire causes heat, for example, simply

---

[21] There is some reason to suspect that Descartes holds that cause and effect are linked by logical necessity. Descartes writes in a letter to More, 'no reasons satisfy me even in physics unless they involve that kind of necessity which you call logical or analytic' (i.e., as the editors of CSMK inform us, that whose denial involves a contradiction) (AT v. 275/CSMK 364–5). See also *Principles*, iv. 206 (AT viiiᴀ. 328/CSM i. 290). I take these remarks to indicate that Descartes regards his physics as a deductive system; I am not sure whether one can turn this sort of claim into a claim about causation per se. On the other hand, as I have been arguing, many of Malebranche's arguments can

because fire would not be fire if it did not. Similarly, on Malebranche's view, it would be a contradiction for God to will that *p* and *p* fail to obtain. Malebranche and Hume cannot be accused of conflating two types of necessity when there was only one to begin with.

Having solved this problem, however, we are immediately faced with another. Even if the collapse of causal into logical necessity is intelligible in its context,[22] why is Malebranche so quick to deny that finite relata, and bodies in particular, can be causes? In the passage just quoted, for example, there is no explicit argument for ruling out finite causal relata: it is just supposed to be obvious that nothing but God's will can live up to the necessity criterion. But it would have been anything but obvious to Suárez or Aquinas.

This question might seem a bit of unnecessary mystery-making, since Malebranche's statement of NNC suggests that any cause will have to be omnipotent. If this is Malebranche's point, then it is trivially true that no physical being is a cause (since it lacks a will), and close to trivially true that no finite mind is a cause. I think this suggestion makes Malebranche's argument implausibly weak; at best, it pushes the question back and makes us ask why anyone would agree that only omnipotent beings can be causes. More than this, however, it gets the structure of the argument wrong: it is because a true cause is one that is necessarily connected to its effects that only an omnipotent being can count as such. The 'most dangerous error of the ancients,' Malebranche thinks, is to assign a logically necessary connection to finite beings on their own; the concurrentists absorb enough of this tradition to include finite beings in the total cause of ordinary events. But again, why is Malebranche so sure that it *is* an error, particularly given that so many other philosophers, spread over nearly two thousand years, found some such view quite reasonable?

We should begin by clearing up a possible source of confusion. Malebranche requires that the mind *perceive* the necessary connection. He then seems open to the objection that he is confusing metaphysics and epistemology. Why should our cognitive abilities be a deciding factor for metaphysics? One might object that even if we grant that causal necessity is logical necessity, sufficiently complicated instances of the latter can elude even the most acute minds. And if Malebranche reads 'the mind' as God's mind, he simply pushes the problem back, inviting the

be seen as developing points that are anticipated by, or implicit in, Descartes's works, so it would not affect my interpretation if Descartes did indeed take causation to be logical necessitation.

22 P. J. E. Kail (2007: 87) goes further and argues that the only alternatives are mere regularity and 'absolute' necessity. Hume is not conflating logical and nomological necessity, for the simple reason that there is only one kind of necessity. As Kail puts it, 'If you maintain that there is metaphysically speaking more to the world than mere regularities, then no other modality is available to you other than absolute necessity.' While I am sympathetic to Kail's position, I do not wish to defend it myself here. My point is that, whether or not there *are* alternatives to logical necessity, the moderns do not see them.

objector to ask how Malebranche has epistemic access to the divine mind and its perceptions (if indeed it has any).

But the epistemic element is really innocuous.[23] In order to work, the objection has to appeal to undetected logically necessary connections. So, even though a and c are not (perceptibly) necessarily connected, there is some intermediate chain of causes and effects $b_1 - b_n$ that are. Malebranche can now respond that, *whatever* one fills in for $b_1 - b_n$, the objector must claim that it is logically impossible for the relation to fail to hold between each pair. And now we simply run NNC on these two events. In short, the problem cannot be the merely epistemic one of locating logical necessities that might after all be there; finite relata are simply not the right sorts of things to serve as truly causal relata.

But why? Malebranche does not tell us. It is not just the reply I have constructed for him but his typically curt dismissal of even the epistemic possibility of a necessary connection between finite relata that suggests he takes it to involve a category mistake. I think we can reconstruct an account of this part of his argument if we recall some of the lessons of the little souls argument.

There, we saw that the chief problem in attributing causal powers to bodies is that it requires us to 'paint the world with the mind's colours.' Powers are directed at non-actual states of affairs and intrinsically point to their actualization. By making explicit what is to count as a genuine cause, NNC lets us take up this issue from the other end: once we see what a true cause requires, we shall see that there is no way in principle for bodies to serve as causes.

We can begin by locating the precise difference between God's will and a given corporeal event or object, say, the striking of a match. Why is it that the match cannot logically necessitate its effect? The chief problem is that the match (or the striking of it) does not, as the little souls argument emphasizes, have any intrinsic connection to its putative effect. By contrast, there is a connection between God's will and its effects that physical events simply cannot have. For a divine volition includes its effect in the sense that that effect is specified as the *content* of the volition. When God wills that this chair move, the two events are linked not by the mere sequence God's volition–chair moving, but by God's volition *with this particular content* and the realization of that volitional content.

We can now state the argument: something is of the right kind to serve as a cause just in case it includes its effect in its content; but only volitions are of the right logical type to do so, because they necessarily include p-volitions; and a p-volition (say, that this chair exist in a given location) *just is* the volition's effect. It is this identity that preserves the necessity: the claim that 'if God wills that p, then p' is necessary because it is analytic; it is analytic not just because God is omnipotent but because the contents of his volitions are identical with their effects. Thus, the cognitive model, suggested by the little souls argument, has a role to play in NNC as well.

---

[23] I owe this point to Steven Nadler.

This is a difficult point, but we can come at it from another angle. The logically necessary connection between cause and effect requires that they be linked in the right way. But how can a physical event, described in a non-question-begging way, point to or be linked with an effect in any way at all? Events described in mechanical terms are not internally connected to their putative effects. For example, 'The ball is dropped from the tower' and 'The ball hits the ground' do not in any sense include or make reference to each other.[24] Only intentionality has this feature of directedness. Thus, the will is perfectly and uniquely suited to play the role of cause. Unlike a bare object or event, a volition can be directed at a distinct state of affairs, simply by including that state of affairs as its propositional content. This, I think, is why Malebranche finds the notion that finite relata could be *causal* relata so obviously muddle-headed.

As we have seen, the dominant view held by Malebranche's opponents is not the thoroughly 'pagan' one that takes finite objects to be autonomous agents but the concurrentist view that includes God in the total cause of any effect. NNC applies equally well to secondary efficient causes taken on their own, since even on the scholastic view, there is no necessary connection between these and any states of affairs. God's concurrence is required. But it also applies to the scholastics' total cause. For the necessary connection here is grounded *both* in God's activity and in the power of the created being. But God's activity is not directed simply at a future state of affairs as such, as it is on Malebranche's view; instead, God works through a created power. The directedness of the total cause, then, must come in part from that created power itself. And this is what Malebranche challenges.[25]

To sum up: Malebranche accepts the scholastic requirement of *esse-ad*; a cause must somehow be intrinsically directed at its effect. But like Descartes, he finds it impossible to conceive how physical objects could have this feature. This, of course, is intimately connected with the abandonment of the Aristotelian ontology. Once a broadly mechanical view is in place, it becomes hard to see how bodies could be causes. Malebranche instead meets the intentionality requirement by ascribing causal power only to the one kind of thing that *can* be directed at non-actual states of affairs: the mind.

If I am right about the role of intentionality in NNC, one might well wonder why Malebranche does not bother to make this more explicit. Given the otherwise mysterious nature of the argument, any interpretation of it will have to answer some such question, as any interpretation will need to go beyond the materials provided by Malebranche's curt statements of NNC. For my part, I think

---

[24] Of course, one could describe these events in such a way as to make them related: consider 'The ball is dropped from the tower to the ground.' But this is surely gerrymandering.

[25] Malebranche does, in fact, distinguish the two views (Aristotle's and the scholastics'). *SAT* vi. ii. 3 is directed at the ancients, while *SAT* E XV is directed at the doctrine of secondary causes. But the latter preserves enough of the former (especially the notions of nature and power) to be defeated by the same lines of argument.

that, given the prominent role of his own version of the little souls argument, Malebranche might well take it for granted that his readers have absorbed the lesson of that argument: a real cause would have to be directed at its effect. And in the course of stating that argument, he makes it quite clear that no physical object or event can be so directed. One upshot of the analysis of Malebranche's suite of arguments I am in the course of offering is that each argument is less a stand-alone, one-off attempt at undermining the scholastic picture than a node on a web of interrelated lines of thought.[26]

I do not mean to underestimate the differences between Malebranchean and scholastic analyses of causation. The logical necessity of true causal claims for the scholastics is grounded partly in creatures; for Malebranche, its source is the divine will. This makes for quite a number of epistemic differences as well: an occasionalist's view of science will not be classificatory in the way Aristotelian science is, nor will it seek to investigate the natures of created beings (see below, Chapter 12). For all causal power now resides in God, and to say that an event was caused by God is to say something that applies to *all* events, and hence is uninformative. None of this, however, detracts from the point of agreement I have located. For the cognitive model is the immediate offspring of the *esse-ad* requirement.

We can now step back from the details of the argument and consider how a concurrentist might respond. As I have shown, NNC takes as its target the conception of causation as logical necessitation endorsed by concurrentism. If an object has the requisite powers, and God concurs, it cannot but produce its characteristic effect when in the presence of objects with the relevant passive powers. It is, for example, impossible that these conditions be satisfied and fire fail to burn human flesh; anything that failed to do so would, for that very reason, not *be* fire. What response could such a philosopher make to Malebranche's claim that it is conceivable, and hence possible, that any event follow any other? The Aristotelian claims that a thing's substantial form is either partly constituted by or logically entails its powers. We must be careful, then, how to describe what it is we are conceiving. When I conceive of myself walking unharmed through a wall of flame, I am conceiving of either a situation in which God does not concur with the fire's power, or in which what I walk through, although it has some of the same superficial properties, is not fire after all. The forms of created

---

[26] One might also wonder why Malebranche does not deploy NNC in answering Bernard de Fontenelle's objection, which Malebranche has Theotimus voice in the *Dialogues*. In Dialogue 7, section 12, Theotimus asks what will happen, given matter's impenetrability, if two objects collide before God has established the laws of motion. This suggests that impenetrability is necessarily connected to motion, even if it does not fully determine the direction or speed of that motion. Malebranche's response is disappointing (see Nadler 2000, who takes a similarly dim view); he simply says that when bodies *a* and *b* collide, God must then decide which laws of motion to set down. Thus, not only does Malebranche not invoke NNC, or any of his other arguments against secondary causes, he sidesteps the whole issue.

things, then, create a network of logically necessary connections. One is not in a position to know these connections unless one has fully grasped the essences of both agents and patients, however, and so many states of affairs will *seem* conceivable, and hence possible, when they are not.

Evaluating the strength of this reply is one of the tasks of the rest of the book. By way of foreshadowing, I might point out here that much will turn on the nature of what one takes as the causal relata. For the Aristotelian, the relata include full-blooded forms, which just are or entail powers. With a robust Aristotelian ontology, it is easy to see how one might argue that Malebranche's criterion of logical necessity is satisfied: if we knew all of the relevant forms, we would also know precisely what would and indeed must happen, given God's concurrence, since this is fixed by what amounts to a set of analytic truths. For Malebranche, by contrast, body–body interaction, if it obtained, would have to connect the bare-bones qualities of extension—size, shape, and motion, which are themselves reducible to relations of distance—and nothing else. If we gave Malebranche his ontology, it would be correspondingly harder to resist NNC, for the Aristotelian reply is not available when talk of powers and forms has gone by the boards. In Chapter 14 below, I consider whether a mechanist like Régis who adopts key notions of the concurrentist framework is able to resist the argument. And throughout Parts II and III, I consider how various philosophers reply, or could reply, to NNC. Keeping NNC and the cognitive model in view should enable us to discover much about the nature of the necessity these figures think is involved in causal transactions.

10.6 THE EPISTEMIC ARGUMENT

Understanding the cognitive model helps illuminate Malebranche's otherwise mysterious argument against the claim that finite minds can be causes. We know that minds, according to Malebranche, are at least of the right ontological type; still, lacking omnipotence, they also lack a necessary connection with their effects. For any instance of a finite volition and its putative effect, we can always conceive of the former without the latter. This shows that the requirement of logical necessity is not fulfilled (*SAT* vi. ii. 3: 450). Given Malebranche's adoption of the scholastics' analysis of causation as logical necessitation, NNC alone is enough to show that no finite mind is a cause. Thus, meeting the intentionality requirement is a necessary, but not a sufficient, condition for causal power.

But there is a deeper issue here. For Malebranche does not rely solely on NNC to show the inefficacy of finite minds; he also thinks that our ignorance of the neurophysiological facts prevents our will from being a cause. As we shall see, this argument is another manifestation of the requirement that cause and effect be linked by intentionality.

In what we might call 'the epistemic argument,' Malebranche asks,

Can one do, can one even will what one does not know how to do? Can one will that the animal spirits expand in certain muscles, without knowing whether one has such spirits and muscles? One can will to move the fingers, because one sees and one knows that one has them. But can one will to impel spirits that one does not see, and of which one has no knowledge? Can one move them into muscles equally unknown, by means of nerve channels equally invisible; and can one choose promptly and without fail that which corresponds to the finger one wants to move?   (*OC* x. 62)[27]

The chief difficulty with finite minds as causes is their lack, not of omnipotence, but of omniscience. As this text makes clear, Malebranche argues that, for my will truly to be a cause, I would have to know much more about the causal sequence involved in moving my arm if I were to bring it about that my arm moves. (Once again, there is a very similar argument in Hume.)[28]

Let us first distinguish between volitions whose contents are identical with their immediate effects and those that are not. Call the latter 'chain volitions,' i.e., volitions whose propositional contents the subject can only bring about by setting a chain of further events into motion. Thus, willing that my car start is a chain volition, since I can only bring it about by turning the key, which in turns sends an electrical impulse down the steering column, and so on. In a chain volition, one wills the outcome of a chain of events.[29]

Now, Malebranche can admit that there is in principle no problem with seeing how a non-chain volition achieves its effect. The problem is that all bodily volitions are chain volitions. When I will to move my arm, the content of this volition seems to be identical with its effect, my arm's moving. But the physiology shows that this is not the case: the immediate effect of my volition cannot be the movement of my arm, since there are necessary intervening events, such as the motion of the animal spirits. So despite appearances, even the simplest voluntary bodily act we can imagine requires a chain volition. And an efficacious volition must be directed at its *immediate* effect, given the cognitive model. Absent the identity between p-volition and effect, the volition cannot be efficacious.

We can now reconstruct Malebranche's argument:

(1) An effective volition is either a chain volition or not.

(2) Willing to move our bodies is a chain volition.

(3) In an efficacious chain volition, at least the first member of the chain must be included in the content of the volition.

---

[27] *Meditations chrétiennes*, VI. 11, trans. Nadler (in Nadler 2000: 122). See also *SAT* E XV: 671.
[28] See *E* 7 and McCracken (1983: 258 ff.).
[29] One can in an attenuated sense be said to will the chain of events itself. For example, in willing the car to start, I of course will the electrical impulse to travel through the wires to the starter, but I don't (or don't necessarily) will those events *as such*.

(4) We do not know what this member is in the present case. Thus,

(5) None of our bodily volitions is efficacious.[30]

The trick is turned by premise 3. For the most natural story here is that the volition sets in motion a series of events that issues in the motion of the arm; why should we assume that the crucial first element must be included in the content of the volition?

It is common to defend (3) by pointing to Malebranche's doctrine of blind will. 'The will is a blind power, which can proceed only towards things the understanding represents to it' (*SAT* i. i. 2: 5). Both Radner and Nadler suggest that, since the will can only achieve that which the understanding presents to it, the mind cannot move the body.[31] But this way of putting matters already brings out the weakness of their reconstruction. In fact, the denial of blind will goes no distance at all toward justifying (3). Malebranche's point about the will is surely sound: one cannot will what one cannot conceive. But of course we can conceive of our arm's moving, even if we have no idea of the intervening causal chain.

To understand the epistemic argument we need to invoke the point I have been pushing toward all along: Malebranche requires that causes and effects be linked by intentional content. Now, in the case of chain volitions, the requisite link does not obtain. For what the physiology shows us is that the connection is not *volition–arm moving*, but *volition–brain event x–etc.–arm moving*. And without including the brain event in the content of the volition, that volition cannot be efficacious simply because the p-volition and the alleged effect are not identical.[32] Willing to move one's arm is rather like willing one's car to start without knowing that one must turn the key.

We saw above in the context of the little souls argument that Malebranche takes this requirement of intentional connection to hold across the board. Entertaining the notion that Malebranche might have imposed an epistemic criterion on causes in general, and not merely minds, Nadler observes that 'it seems to be a category mistake to extend the epistemic condition to causation by corporeal agents, such as fire and stones.'[33] But once we see the need for a tie between cause and effect that only intentionality can supply, we also see that this extension is no category mistake; the category mistake is committed by those who claim that finite beings can be causes.

---

[30] It is important to see that the argument is a *reductio*. That is, it begins by taking for granted that there is a genuine causal sequence, initiated by the volition of a finite mind, that issues in a physical event. NNC has already shown that no such connection could obtain between the bodily events Malebranche has in mind.

[31] See, e.g., Radner (1978: 18) and Nadler (2000: 122–3).

[32] Note that I am not claiming that meeting the intentionality requirement is sufficient to count as a cause; the intentionality requirement is a necessary, but not sufficient, condition for causal power. For, as we have seen, Malebranche also holds the other aspect of the scholastic view, namely, that cause and effect be connected by logical necessity, which of course does not obtain in the case of finite minds.   [33] Nadler (2000: 125).

There is a further strand to the epistemic argument that will be important when we come to consider Malebranche's position on laws below. For it is not merely their failure to achieve the identity between the propositional content of a cause and the effect itself that prevents minds from being causes; it is their inability both to establish and to follow the laws that govern mind–body behavior. Consider this passage from the *Elucidations*:

how is it conceivable that the soul should move the body? Our arm, for example, is moved only because spirits swell certain of the muscles composing it. Now, in order for the motions that the soul impresses on the spirits in the brain to be communicable to those in the nerves, and thence to others in the muscles of our arm, the soul's volitions must multiply or change proportionately to the almost infinite collisions or impacts that would occur in the particles composing the spirits; for bodies cannot by themselves move those they meet . . . But this is inconceivable, unless we allow the soul an infinite number of volitions for the least movement of the body, because in order to move it, an infinite number of communications of motion must take place. For, in short, since the soul is a particular cause and cannot know exactly the size and agitation of an infinite number of particles that collide with each other when the spirits are in the muscles, it could establish neither a general law of the communication of motion, nor follow it exactly had it established it. Thus, it is evident that the soul could not move its arm, even if it had the power of determining the motion of the animal spirits in the brain. These things are too clear to pause any longer over them.   (*SAT* E XV: 671)

Mental causation requires that the mind be able to legislate the laws that govern mind–body union, and then remember them for future reference. That is, to move my arm, it would not be enough for me to bring about the causal sequence; I would have also to decree that this particular causal sequence is *the* sequence that brings it about, and then remember this in future whenever I wanted to move my arm. I would have to set down the rules of the game and then be able to follow them. But this game, Malebranche argues, is too complicated. Here again we see the Cartesian top-down conception of laws: if there is a genuine causal sequence, it must be governed by laws, and these laws require a lawgiver. It is not enough for my will merely to 'obey' these laws in the attenuated sense of acting in accordance to them. (This line of thought will become important in section 11.2 below.) In this sense, to accord ourselves causal power is to make ourselves into tiny gods, each able to decree a law of nature.

I take these arguments to constitute the gravamen of Malebranche's case against the attribution of causal powers to bodies and finite minds. At various points in what follows, each of them will recur, as will Descartes's arguments against the scholastic notion of power. For it is only against the background of these attacks that we can clearly see the notions of power and cause taking shape in philosophers like Régis, Boyle, and Locke.

Occasionalism is a highly counter-intuitive doctrine, and Malebranche surely thinks that he has powerful considerations to advance on its behalf. If these considerations are left opaque, his view cannot but seem a historical aberration. But if I am right, the key element of Malebranche's dialectical strategy—the cognitive model of causation—falls into place, and the appeal of his arguments, at least in their proper intellectual context, becomes clear. Why would someone think that attributing powers to bodies requires thinking of them as possessed of little minds, with little wills of their own? Well, powers are supposed to be the sorts of things that of their own nature tie an event to its effect. But the only plausible candidate for such a tie is the relation of intentionality, as captured by p-volitions. Why should we think that the first step in a chain volition must be included in the content of that volition? Again, only because a cause must be p-volitionally identical to its effect. Where this tie of intentionality is absent, all events are indeed 'entirely loose and separate,' as Hume was to write. Malebranche inspired Hume's claim that 'Solidity, extension, motion; these qualities are all compleat in themselves, and never point out any other event which may result from them' (*E* 7. 8). For unlike Aristotelian powers (or Malebranchean volitions), these mechanical qualities are not of the right ontological type to be connected to their effects.

11

Laws and Divine Volitions

11.1 THE CONTENT OF DIVINE VOLITIONS

Thus far, I have said little about Malebranche's own positive view. How do the laws of motion function, in the context of occasionalism? And how are we to understand the divine volitions that contain or execute these laws? Let us begin with this second question, in the hope that it will help us answer the first.

Commentators have split roughly along the following lines.[1] What we might call the Leibniz–Nadler reading attributes to Malebranche's God fully particular p-volitions and thus makes him causally active at each and every moment.[2] If body x moves to place b at t, then God must have willed *that x move to b at t.* Thus, Malebranche's God must, Leibniz thinks, constantly perform miracles.[3]

It is important not to be misled by the suspicion that this interpretation is inconsistent with God's timeless nature. God's p-volitions can be temporally indexed, even though his volitions themselves are not. And of course one can capture the timelessness by just piling all of the particularized p-volitions, fully temporally indexed, into a single super-volition that God wills from eternity. Thus, Leibniz's complaint that Malebranche's God is a busybody who constantly

---

[1] I should point out that the controversy at issue here—determinate–indeterminate volitions—is orthogonal to the distinction Malebranche draws between general and particular volitions. Briefly, a general volition is one that is at least in accordance with the laws of nature, while particular volitions are the causes of miracles. I say a bit more about this below.   [2] See Nadler (1993*b*).

[3] In fact, I suspect Leibniz also has a distinct point to make by accusing Malebranche of forcing God to perform continual miracles. It is not merely that the occasionalists' God has to intervene continually in the course of nature; rather, at least one of Leibniz's points is that the occasionalist is, as it were, cheating, by not producing an explanation of phenomena that appeals to the natures of substances. In 'A New System of Nature,' Leibniz writes, 'It is quite true that, speaking with metaphysical rigor, there is no real influence of one created substance on another, and that all things, with all their reality, are continually produced by the power of God. But in solving problems it is not sufficient to make use of the general cause and to invoke what is called a *Deus ex machina*. For when one does that without giving any other explanation derived from the order of secondary causes, it is, properly speaking, having recourse to a miracle. In philosophy we must try to give reasons by showing how things are brought about by divine wisdom, but in conformity with the notion of the subject in question' (AG 143). It seems to me that Leibniz's real point here is nothing to do with the ubiquity of God's activity, on the occasionalist view, but with their failure to explain phenomena by means of the subjects involved in producing them. Thus, Leibniz presents his own notion of an individual substance, which develops in virtue of its own active power, and which, unlike the scholastics' powers, needs no external stimulus to reduce it to act.

meddles in the world by means of a 'perpetual miracle' is only partly accurate. Although God's activity is of course ubiquitous, God need not constantly adjust his activities in light of the states of the world. Instead, God might simply will one (indefinitely long) p-volition eternally, each of whose propositional constituents is temporally indexed.

On Nicholas Jolley's reading, by contrast, God is 'limited to willing the initial conditions of the universe and the laws of nature.'[4] In allying himself with Antoine Arnauld, however, Jolley seems to have made the mistake I have just sketched out. For the passage he quotes from Arnauld as evidence that he subscribes to Jolley's own 'minimalist' view is inconclusive. Objecting to Leibniz's characterization of Malebranche, Arnauld says that,

Those who maintain that my will is the occasional cause of the movement of my arm and that God is its real cause do not claim that God does this in time by a new act of will each time that I wish to raise my arm, but by that single act of the eternal will by which he has willed to do everything which he has foreseen it will be necessary to do, in order that the universe might be such as he has decided it ought to be.[5]

Everything Arnauld says here is consistent with, and indeed points towards, the essential claim of the Leibniz–Nadler view, which holds that God's p-volitions are fully determinate. Whether God needs one volition or many to accomplish what he wants is beside the point. How many volitions it takes to lick to the Tootsie Roll center of a Tootsie Pop is irrelevant to the central issue which concerns, not the number of acts of will, but their content. That God requires only one act of will tells us exactly nothing about the content of that act.

It is well to get this out of the way at the start, for I suspect that any dispute about the number of God's volitions (as opposed to their content) would be very difficult to settle, even within the context of Malebranche's view. For we would need some way to individuate acts of will over and above their contents, since, as I have argued, there is no reason to doubt that a single divine volitional act can be directed at multiple states of affairs. It is questionable whether we have a grip on how to individuate volitions otherwise, or what this even means in the context of our own mental lives, much less God's.

Whatever its historical antecedents, Jolley's view gains plausibility by appealing to parsimony. Malebranche's God never acts except in the simplest possible way (*SATE* XV: 663). Now, Malebranche, like Descartes in the *Principles*, treats the laws of nature as causally active. If the initial conditions of the universe plus the laws of nature are sufficient to determine the course of events, Jolley argues, then God's p-volitions need not refer to each individual body as such. The contents of God's nomological volitions are conditionals to the effect that if a body has such-and-such set of modes, it will be moved (or not) in this or

---

4 Jolley (2002: 246). See also Clarke (1989: 121 ff.), who endorses the Arnauld interpretation.
5 Arnauld to Leibniz, Mar. 4, 1687, quoted in Jolley (2002: 246).

that direction and speed. Jolley's God, then, is responsible for all physical events only in the sense that he wills the initial state of the universe plus the laws of nature.

Jolley claims that his view alone captures the causal role Malebranche gives to laws of nature. On the Leibniz–Nadler reading, by contrast, it seems that these laws will either be mere statements of regularities that describe God's behavior or general principles God keeps in mind when re-creating the world. In neither case does it make much sense to call these laws 'causes.' By contrast, if Jolley is right and the laws are part of God's p-volitions, then they can meaningfully be said to be causes.[6]

There are independent and decisive reasons to reject Jolley's interpretation. As he is well aware, it seems in direct conflict with the divine concursus argument. For a key premise of that argument is that God's p-volitions must be fully determinate. Jolley initially suggests that the argument might be merely *ad hominem*, directed at other Cartesians. He grants that there is no evidence that Malebranche intends it in this way, but matters are worse for this proposal, since other key arguments, such as NNC, rely on some of the same premises, particularly the conceivability principle.

Jolley abandons this suggestion and goes on to attempt a reconciliation of his reading with the divine concursus argument. He claims that 'the doctrine of continuous creation can be reductively analyzed in terms of God's efficacious general volitions':

> To say that all of the billiard ball's states depend on God as a causally sufficient condition is to say that they can all be genuinely explained in terms of God's general volitions (the laws of physics) and the initial conditions which he wills. The doctrine of continuous creation is thus very far from requiring particular discrete volitions corresponding to each state of a creature.[7]

This does indeed capture a notion of dependence that would be acceptable to Malebranche. But it goes no distance at all toward allowing us to understand how God can will to create the billiard ball and all of its states in a given location and time without having a fully determinate p-volition. In other words, Jolley has given us a way to think of the billiard ball's states as the effects of God's activity, whereas what is needed is a way to think of the contents of God's *will* as leaving indeterminate the location of the ball. Since it is impossible for the ball to exist without possessing a location, it is also inconceivable and hence unwillable. The dependence of the ball's states on God is not the issue.

What is more, the principle of parsimony cuts both ways. One might think, as Jolley does, that if God can get away with just willing the laws of nature and initial conditions, this is what he *must* do, given his perfection. But it

---

[6] I deal with this line of thought in the next chapter.   [7] Jolley (2002: 251).

would be equally plausible to say that since we know from the divine concursus argument that God must will the existence of bodies and their locations, it would be superfluous for him to engage in a further act of will, namely, one whose contents included the laws of nature. On the Leibniz–Nadler view, laws supervene on God's acts of will; they have no status above and beyond this will.

Finally, it seems to me that Jolley's view is anachronistic in so far as it presupposes that once God wills a law of nature, it continues to 'govern' events on its own, as an autonomous feature of the universe (albeit one with a source in God). It is not enough simply to will conditional claims; one must also bring it about that their consequents come to pass when their antecedents are fulfilled. To suppose otherwise is to suppose, in Cudworth's mocking phrase, that the laws of nature could 'execute themselves.'

Malebranche does, of course, draw his own distinction between general and particular volitions, and this can seem to lend strength to the Jolley view. Malebranche claims that 'God must not act as do particular causes, which have particular volitions for everything they do' (*SAT*E XV: 667). The context of this remark—the doctrine of predestination—is a vital clue that what is at issue is the problem of evil. This is more explicit in the *Treatise on Nature and Grace*, where Malebranche considers the case of an infant who is born 'with a monstrous and useless head growing from his breast that makes him wretched.' This fact, Malebranche claims, obtains 'not because God has willed these things by particular volitions, but rather because He has established the laws of the communication of motions, of which these effects are the necessary consequence.'[8] The distinction Malebranche draws here is not between volitions with fully specified contents and volitions that specify only the laws of nature. Rather, it is the difference between willing an effect for its own sake and willing it only as a consequence of further facts. The distinction thus embodies the doctrine of double effect. Although any effect, including deformities and damnation, is the effect of God's will, God does not will all effects for their own sake. The question is not the content of the volition but the intent behind it. Just as it would be disingenuous to pretend that a doctor performing an abortion to save a mother's life does not will that the fetus be killed, it would be silly to think of God as not willing that little Jimmy be squashed by a falling boulder. But it would be less silly, or so Malebranche thinks, to claim that neither God nor the doctor willed their effects *tout court*, i.e., would have willed them had the situation been otherwise.

If all of this is right, then, there is good reason to think that the contents of divine volitions must be fully determinate and temporally indexed. But what, then, becomes of laws of nature?

---

[8] In Nadler (1992: 260).

11.2 THE PROBLEM OF EFFICACIOUS LAWS

The problem of the nature and status of Malebranche's laws can be made more acute if we consider Malebranche's claim that laws are causally efficacious. It is, on the surface, very hard to see how a proposition, whatever one's ontology, could act on the world.[9] We should begin by noting an obvious constraint on any interpretation of this issue, namely, the status of laws as dependent on God. There is no sense to be made of the idea that God wills the laws of nature and bodies behave according to them on their own, as it were. The reasoning here is just the same as it was with regard to Descartes: laws lack the kind of independent ontological status that would be required if they were to be autonomous causal agents. So whatever picture we end up with, it has to be the case that laws are identified, one way or another, with God's volitions and not with their effects.

It seems clear enough that Malebranche is committed to the causal efficacy of laws:

All natural forces are therefore nothing but the will of God, which is always efficacious. God created the world because He willed it: 'Dixit, & facta sunt' (Ps. 32:9) and He moves all things, and thus produces all the effects that we see happening, because He also willed certain laws according to which motion is communicated upon the collision of bodies; and because these laws are efficacious, they act, whereas bodies cannot act. There are therefore no forces, powers, or true causes in the material world... (*SAT* vi. ii. 3: 449; cf. *OC* x. 54)

What could be going on here? Malebranche clearly claims that laws act, and in the same breath that God moves all things. On the very next page, he says that 'it is only God who is the true cause' (*SAT* vi. ii. 3: 450). In fact, the question of how Malebranche conceives the content and causal efficacy of laws is significantly more complicated than has been realized. I shall argue that Malebranche's texts suggest two distinct analyses of law statements, only one of which is consistent with attributing causal efficacy to them.[10]

First, Malebranche claims that laws are *occasional* causes. In the *Treatise on Morality*, Malebranche also says that the laws of the union of mind and body are 'the occasional causes of all these confused and lively sentiments' (*OC* xi. 116). These laws are 'efficacious in virtue of the action of divine volitions which alone are capable of acting on me' (*OC* xi. 117). At least these laws, then, are not genuine causes after all; like the modes of bodies, they are merely prompts for God to act. This is the force of calling laws 'occasional' causes.

The picture this suggests is reasonably clear: God, so to speak, looks at the state of the world, considers the laws he has willed, and arrives at a new volition

---

[9] Jolley's ingenious answer appeals to a parallel with efficacious ideas; see Jolley (2002: 256).
[10] For more on the two analyses of laws—conditional and summary—see above, sect. 1.1.

that produces objects in locations consonant with these facts.[11] The laws, then, would presumably be conditional statements that God eternally wills. But, contra Jolley, they would not themselves serve as real causes, even conjoined with facts about the initial state of the universe. These conditional statements would be demoted to the level of occasional causes.

This analysis gains independent support from reflecting on the necessary connection requirement. Just as Jolley holds, laws on this model are general statements to the effect that if *x* happens, *y* will happen. But, so construed, laws can *only* be occasional causes.[12] For as Jolley notes, laws are not necessarily connected with any particular events. (God could, after all, will that the conservation of motion hold in a world where no events satisfy its antecedents.) Jolley's answer is that God's nomic volitions must be understood as including the initial states of the universe, which together determine events. But this must seem ad hoc; if Jolley is willing to bundle initial conditions into God's volitions, why not allow the particularist reading, full stop? So the *Treatise* account is not a mere aberration, but a view with roots in the rest of Malebranche's position. Thus, the conditional analysis generates what we might call the 'weak' (inefficacious) conception of law.

In principle, I suppose, there's no reason why the *Treatise* model could not be reconciled with the texts from *The Search After Truth*. We would have to assume that Malebranche intends us to understand that laws are causes only in the attenuated sense he extends to natural bodies. But this is not fully convincing; it is hard to know why Malebranche would not be more careful in a section where he is explicitly concerned to distinguish real from natural causes.

Thus, we need to consider an alternative. Malebranche sometimes suggests that the laws of motion are to be reduced to God's fully particularized (non-conditional) volitions. We know from the divine concursus argument that God must will each particular fact as such. In the *Dialogues*, Malebranche states this explicitly when he writes that the laws that govern mind–body relations '*are nothing but* the continual and always efficacious volitions of the Creator.'[13] It is enough if God will the events that underwrite the conditionals in question, for once he efficaciously wills a given effect, this entails that a law of nature has been established. That is, once God brings about a given effect, it follows that a law has been established, just because that law is nothing but a statement of the sequence *state of affairs q at t – God wills that p at t′ – state of affairs p at t″*.

---

[11] This is only a rough illustration, not to be taken literally. I have explained above that Malebranche's God does not act in time.

[12] See above, Ch. 7. There, I argued that Cartesian laws cannot be genuine efficient causes, since it is, in the end, only God's will that is efficacious. Nevertheless, I suggested they played some of the traditional roles assigned to secondary causes. It is important to keep in mind that Malebranche is openly hostile to the doctrine of primary and secondary causes and, unlike Descartes, has no wish to preserve even a terminological similarity with it.

[13] Dialogue XII, *OC* xii. 279; my emphasis. Oddly enough, Jolley himself quotes this passage (2002: 252), though I think it fairly clearly tells against his view.

In contrast to the conditional model, this 'summary' model lets us make sense of laws as genuinely causally active. Unlike conditional statements, they meet the necessary connection requirement. If laws are reduced to God's activity in the world, then of course they will be necessarily connected with the events in question. For a law just is one of God's volitions, and these individual volitions are necessarily efficacious, given their source in an omnipotent being. This must be the model Malebranche has in mind when he claims that 'while nature remains as it is, i.e., while the laws of the communication of motion remain the same, it is a contradiction that fire should not burn or separate the parts of certain bodies' (*SAT*E XV: 662). The contradiction arises only if those laws are God's (fully particular) volitions.

To sum up our results so far: Malebranche's corpus offers two analyses of law. On the weak version, laws can be taken as general conditional statements. But for that very reason (among others), they can only be occasional, and not real, causes. On the second analysis, laws are reduced to the individual volitions by which God conserves objects. Laws then can indeed have causal efficacy, but only because these laws are not really laws at all but convenient ways of talking about God's behavior.

I think we can see both analyses operating side by side in a single sentence of the epistemic argument I flagged above (section 10.6). Malebranche argues that a finite mind 'could establish neither a general law of the communication of motion, nor follow it exactly had it established it' (*SAT*E XV: 671). But why would causing an event require one to 'establish' a law of nature? Imagine Adam in the garden of Eden. And imagine, *per impossibile*, that God has endowed Adam with an efficacious will. Malebranche holds that when Adam wills to move his arm for the first time, he establishes a law of nature that did not previously exist. But there seems to be no motivation for this move; why not just think that he's following a law that is already there?

This makes sense only if laws are indirect ways of talking about actual causes. There is no difference in kind between a singular efficacious volition and a set of them. So even a single volition amounts to a law. Adam establishes a law only in the sense that he performs an efficacious volition. The first part of this sub-argument, then, requires that laws be understood as the 'strong' model requires.

Without this identification, I can see no way to make sense of this epicycle of the knowledge argument. Unfortunately, it seems to undermine the very next move of Malebranche's argument. For Malebranche suggests that Adam's establishment of the law of nature requires that he follow that law when he efficaciously wills the same act in the future. But this makes sense only on the first, 'weak' version. That is, only if we conceive laws as (inefficacious) conditional claims can we think of Adam in any sense establishing a law that he would in the future have to follow. No doubt Malebranche is using this model to make the same kind of point he made in the 'little souls' argument: the orderliness of the world depends on the omniscience of a divine being. If our minds were indeed causes, their lack of omniscience (especially as applied to their own capacities of memory) would make the world a chaotic place. So although Malebranche does oscillate between these two analyses of law, he has a point to make by using each.

Although the two analyses are intensionally distinct, and seem to come apart when we consider a finite being like Adam as a cause, there is good reason to think that, as applied to God, they are extensionally equivalent. What unites them, I think, is God's omniscience. In the context of omniscience there is no cash value to a distinction between the fully spelt-out p-volitions of the second analysis and the conditional p-volitions of the first. If I know everything, then there's no difference between my willing that if one thing happens, another will happen, and willing everything that will in fact happen as such.[14] Preferring the latter formulation is merely a way to acknowledge that God's volitions must be fully determinate; but there is no way they could fail to be, since everything is (cognitively) present to God.

It makes sense, then, to take the distinction between a law of nature as a general, conditional statement and as a set of individual, fully particular volitions as a purely conceptual one. In the context of explanation, it seems obvious that we will be best served by thinking of laws under the weak description. A fully worked-out mechanics, then, would simply be an exhaustive statement of all of these conditionals. But that does not mean that God wills those conditionals as such, as opposed to what makes them true.

[14] I owe this point to Andrew Chignell.

12

Causation and Explanation

I have not made such heavy weather of Malebranche's conception of laws for its own sake. What we might think of as the trial separation between causation and explanation in Descartes's work issues in permanent divorce in Malebranche's. His severing this tie has a profound influence on later thinkers such as Berkeley and marks the first clear articulation of the independence of natural philosophy from ontology.[1]

Malebranche, as we have seen, has two distinct analyses of laws. At some times, he treats laws as conditional statements; at others, he collapses them into the fully particular p-volitions by which God continually re-creates the world. In terms of explanation, the first of these must take precedence; and for all the reasons we have seen, laws on this account cannot be causally efficacious. This, I believe, is the kind of picture Malebranche has in mind when he claims that 'extension by itself with the properties everyone attributes to it is sufficient to explain all natural effects.'[2] When considering laws as conditional statements, the only further information one would require to achieve explanation would concern the instantiated modes of extended substance.

It is the first analysis of laws that provides them with explanatory power. This fits well with Malebranche's frequent injunctions that the true cause, namely, God, is not to be invoked to explain any particular finite event. '[W]e would be ridiculous were we to say, for example, that it is God who dries the roads or freezes the water of rivers' (SATE XV: 662). A natural science directed at uncovering true causes would be extremely easy to master, for all questions under its purview would have the same answer: God. Thus, the weak conception of laws, embodied in the first analysis, is the only one that truly divorces causation from explanation. If we held the second analysis, all explanation would appeal to God's volitions themselves in the most direct possible way.

The clearest articulation of this position, however, comes not from Malebranche but from Berkeley.[3] For Berkeley, natural philosophy is the process

---

[1] A helpful contrast here is with Spinoza, who ties causation and explanation together as tightly as one could wish. (For more on this, see Carraud 2002.) One can then see Spinoza and Malebranche as taking up strands of thought in Descartes and developing them in opposing ways.

[2] Dialogue III, in Nadler (1992: 176).

[3] For evidence of Malebranche's direct influence on Berkeley, chronicled in McCracken (1983), see Berkeley's *Philosophical Commentaries*.

of discovering laws, not causes. 'It is not . . . in fact the business of physics or mechanics to establish efficient causes, but only the rules of impulsions or attractions, and, in a word, the laws of motion, and from the established laws to assign the solution, not the efficient cause, of particular phenomena' (*DM* 35). To borrow an image of Helen Beebee's:[4] suppose God were to list all the facts that together exhaustively characterize the universe throughout its entire history. We, of course, could never come close to fully reading or comprehending this list. So to help us out, God provides an axiomatization, a set of generalizations that allow us to derive some propositions from others. These generalizations are not themselves further truths that have to be included in the list of facts; they are just handy tools. What is more important, they are not the truth-makers of any of the propositions on the list. Now, God always acts in the simplest possible ways, which guarantees that such an axiomatization is possible.

Conceiving of laws as conditionals, Berkeley argues that scientific explanation is nothing more than subsuming a particular phenomenon, or set of phenomena, under a law. This is a sort of functional replacement for the Aristotelian dream of *scientia* as a network of demonstrations. Even though the definitional basis of such demonstrative knowledge has been removed, it remains the case that science is a matter of tracing out connections between particular and general propositions. There is an important difference, however: while we can deduce many propositions of natural science, we cannot demonstrate them, since, as we have seen, demonstration requires deduction from necessary truths. And, as Berkeley notes, we cannot assume that the 'Author of nature always operates uniformly' (*PHK* I. 107). Causation, then, is one thing; explanation, quite another. Thus, with Malebranche and Berkeley, we have left behind the last vestiges of the bottom-up position, and arrived at a self-conscious transformation of the role of natural philosophy.[5]

---

[4] See Beebee (2004).
[5] For a different and quite useful way of mapping out the debate over explanation in the modern period, see Nadler (1998).

13

A Scholastic Mechanism

We have seen how a thoroughgoing commitment to ontological mechanism seems to rule out course-of-nature mechanism: by denying powers to bodies, ontological mechanism has made the top-down conception of laws of nature inevitable. For nothing in those bodies, deprived of Aristotelian powers, can fix, or even help to fix, the course of nature. Further arguments have shown how, given the concept of a law of a nature operative in this period, the top-down conception leads to occasionalism.

It is now time to turn to the other side of the debate. In the remainder of this part of the book and in the next, I am interested in tracing out the influence of key features of the scholastic position, particularly its bottom-up conception of power and its views on the necessity of causal relations. But more than this, I wish to reconstruct and analyze the arguments that led to the revival of these notions and to discover how and with what degree of success some moderns attempt to preserve what was worthwhile in the scholastic view in the context of the radically new ontology of mechanism. Is there some way to combine ontological and course-of-nature mechanism without collapsing into scholastic obscurity? Only by asking this question can we see these other moderns' achievements for what they are. What is more, it is only against this background that we can hope to make sense of Hume's attacks on causal realism.

The traditional *via media* between occasionalism and conservationism among the scholastics was, as we have seen, concurrentism. Although Descartes himself is not plausibly read as a concurrentist, the view was nevertheless to experience a resurgence in the late seventeenth and early eighteenth centuries, at the hands of Pierre-Sylvain Régis (1632–1707).[1] In 1690 Régis publishes his *Cours entier de philosophie, ou, Système général selon les principes de Descartes*; despite its title, it includes many departures from Descartes's own views, as we shall see. Fourteen years later, Régis's *L'Usage de la raison et de la foi* appears, covering many of the same metaphysical and physical questions. Particularly in the latter work, on which I shall focus, Régis melds ontological mechanism and concurrentism.[2]

---

[1] For biographical material on Régis, see Mouy (1934: 135 ff.). Mouy also provides a very brief summary (150–6) of Régis's metaphysics and physics. More useful is Clarke (1989).

[2] Régis's earlier *Système* might well be said to embody a more orthodox Cartesianism, and hence to provide a more top-down picture. Clarke (1989: 150–2) reads Régis this way; note that he

Let us take a step back to reconsider the overall debate. It is natural to wonder why the conception of causation as logical necessity persists in the work of Régis long after its ontological base, namely, Aristotelian powers, had been demolished. But as the cognitive model itself shows, the Aristotelian concept of power is not rejected wholesale but taken up in an entirely new ontological framework.[3] Similarly, Régis can be read as remaking core features of scholasticism in the mechanist image. Thus, in Régis, we find a quite different model of causation, the geometrical model. Régis retains the features of logical necessitation and directedness. Unlike Malebranche, however, he thinks that the bottom-up concurrentist picture can accommodate these in the context of the new ontology.

I shall argue that Régis's position is on the whole superior to those available in the broader context of seventeenth-century Cartesianism. What is more, Régis's own arguments against (and replies to) the occasionalists can tell us much about the dialectic of causation within Cartesianism. And when we turn to his positive views, we shall find him making some moves that parallel those of Locke and Boyle, who were already engaged in making the philosophical world safe for powers.

We can begin by seeing that occasionalism suffers from a serious internal difficulty. The problem comes from admitting that material beings have their own immutable natures while at the same time according all responsibility for their behavior to God. For these natures, however thoroughly stripped by the mechanist ontology, must contribute something, after all, to the course of events. Bernard de Fontenelle initially lodged this objection, which Malebranche attempts to answer in his Seventh Dialogue. Suppose bodies *a* and *b* are in motion and collide. Suppose further that God has not settled any laws of motion at this point. Even so, the nature of matter, specifically, its impenetrability, means that

focuses almost exclusively on the *Système* and not the later *L'Usage*. Régis's earlier work does indeed have a role for laws (and rules) of nature, some of which follow from metaphysical axioms (see esp. *SGP* i. 69 ff., 332 ff.). Thus, for example, the early Régis deduces the inertia principle from the metaphysical axiom that no body can change without being caused to do so by another body. I confess it is still not quite clear to me whether the *Système* in fact commits Régis to a top-down picture. To use a metaphysical axiom in deducing a law of nature is not necessarily to say that that law has nothing to do with the nature of body; indeed, axiom 4, which is in question, follows from the denial of a self-moving principle in bodies. I should take this opportunity to point out again that the contrast between top-down and bottom-up views is not between those views that admit laws of nature and those that do not. Instead, it holds between views that take laws of nature to have their source, if any, in features of the universe other than those of the bodies whose behavior they prescribe and those that do not. That a proponent of the bottom-up picture should speak of laws is in no way incongruous; the real question is the origin and status of those laws. That said, I think it is plausible that in this last respect, Régis's views underwent a change; hence I shall focus on his more mature work, *L'Usage*.

[3] Note that the cognitive model as such does not entail that causes (logically) necessitate their effects (Descartes, as we have seen, holds the cognitive model and yet thinks finite minds can be causes). I mention it here as an instance of a scholastic view that outlives the demise of the Aristotelian ontology.

a and *b* will at least not pass through each other. And this indicates that their natures play some causal role. Malebranche's reply is that, when *a* and *b* collide, God must then make up his mind and institute a law of motion.[4] The collision serves as an occasion which obliges God to distribute his actions in one way or another. But this hardly solves the problem; *why* must God make up his mind? Presumably because *a* and *b* are the kinds of things they are. And this suggests, again, at least a minimal causal role for bodies. Post-Cartesians thus face the same difficulty their scholastic forebears did: how to reconcile God's ubiquitous activity with the natures of created beings. But these inheritors of Descartes's world-view confront this difficulty with a more limited array of choices: occasionalism faces grave problems, as we have seen, while the typical scholastic solution had already been subjected to the withering critiques of Descartes and Malebranche.

To navigate between God's role as the source of being and the role of creatures in determining the course of natural events, Régis remakes concurrentism in the mechanist image. To concur, he says, 'is to join forces with those of another agent, to produce together an effect which could not be produced by either of these forces alone.'[5] Secondary causes are thus insufficient for the production of a given effect; but so is the primary cause, namely, God. But how exactly are secondary causes to be understood?

If we distinguish the role of these secondary causes from that which fills them, we can see that with regard to the former Régis follows scholastic concurrentism quite closely, while he departs from them on the issue of how that role is realized. Closing his attack on the occasionalists, Régis quotes Aquinas with approval:

St Thomas speaks in favor of instrumental causes thus: *The second, instrumental cause*, he says, *does not participate in the action of the principal cause except in so far as by something proper to itself it contributes to the production of the effect of the principal agent; for if the second cause contributed nothing proper to itself, it would become useless, and it would not be necessary to have different instruments for producing different determinate actions.*[6]

Since all change in bodies is a result of motion,[7] an understanding of the natural world must begin there. Descartes had defined motion (in the strict sense) as 'the transfer of one piece of matter, or one body, from the vicinity of the other bodies which are in immediate contact with it, and which are regarded as being at

---

[4] Malebranche (1992: 232–3).

[5] As there is no English translation of any of Régis's works, I give the French in footnotes, and my own translations in the text. '[C]'est joindre les forces à celles d'un autre agent, pour produire ensemble quelque effet, qui ne pourrait être produit si ces forces étaient séparées' (*URF* 951).

[6] 'Voicy comment saint Thomas parle en faveur des causes instrumentales. *La cause seconde instrumentale*, dit-il, *n'a part à l'action de la cause principale qu'entant qu'elle contribue par quelque chose qui luy est propre à la production de l'effet du principal agent; car si elle ne contribue rien qui luy fût propre, son concours deviendroit inutile, et il ne seroit pas necessaire d'avoir des instrumens differens pour produire des actions déterminées*' (*URF* 416; the quotation is from *ST* I q. 45 a. 5).

[7] *URF* 269.

rest, to the vicinity of other bodies' (*Principles*, ii. 25, AT viiiA. 54/CSM i. 233).
As Régis sees matters, Descartes's definition imports a subjective element, in so
far as the transfer of a piece of matter is a transfer only against the background of
a set of bodies that are regarded as being at rest. (Whether this is a fair charge or
not is open to debate, since Descartes struggled to purge his definition of motion
from any such subjectivity.) Régis sees this as a terrible mistake, and announces
that 'We have departed from an eminent modern philosopher.'[8] 'Nothing could
be less well founded than this opinion, since the relation of one body to distant
bodies that one considers as at rest is nothing but a purely extrinsic denomination,
which changes nothing in the bodies that one considers.'[9] Nor is Régis alone
in his complaint. Newton's unpublished *De Gravitatione* is a sustained attack
on the relativity of Cartesian motion.[10] Instead, Régis defines motion as 'the
successive application [impact, collision] of bodies with one another.'[11] (It is not
clear that this is much of an improvement: if anything, Régis has offered us a
cause of motion, but not a definition of motion itself.)

In order to explain how God and creatures concur in producing effects, Régis
isolates two aspects of motion. Motion considered in the mover is nothing but
the will by which God produces 'l'application successive' of bodies. Régis thus
agrees with the occasionalists that matter is inert and that God is the sole efficient
cause of motion taken in this sense, which Régis calls efficient motion or the
moving force. Motion considered in its usual sense, that is, in the object moved,
he calls formal motion.

Formal motion itself can be taken in two senses—substantially and modal-
ly—and here is where the disagreement with the occasionalists emerges. Modal
formal motion is just motion considered as a mode of a body. But Régis also treats
formal motion as a kind of quasi-substance, which is capable of taking on modes
of its own. God, being immutable, immediately produces only substantial formal
motion, which is unchanging with respect to both its nature and its quantity.[12]
To explain why bodies move in the particular ways they do, however, we need to
invoke second efficient causes, in this case, the modes of bodies themselves. Just
as Aquinas's second causes 'particularize and determine' the acts of the primary
cause, so modes of bodies make motion take on the forms that it does. One can

---

[8] 'Nous avons abandonné un Philosophe moderne tres considerable' (*SGP* 302). A marginal
note cites 'Descartes, dans la 2. part de ses principes art. 25.'

[9] 'Il n'y a rien de plus mal fondé que cette pretention, estant tres-constant que la rapport
d'un corps à des corps éloignez qu'on considere comme immobiles, n'est qu'un pure denomination
exterieure qui ne change rien dans les corps où la considere' (*SGP* 302).

[10] Newton (2004). As Andrew Janiak points out in his introduction, *De Gravitatione* shows that
Newton's initial target in defending his realist conception of space was not Leibniz but Descartes.
Disalle (2002: 37) argues that Descartes tacitly endorses the relativistic conception of motion
because it allows him to endorse both the heliostatic and geostatic models of the solar system.

[11] 'l'application successive des corps les uns aux autres' (*URF* 960).

[12] 'Or Dieu ne produit immediatement que la substance de la mouvement formel; car pour les
modes de ce mouvement, ils dépendent immediatement des creatures' (*URF* 296).

think of this on analogy with a quantity of water being driven through a number of differently shaped channels. Although the shape of the channels on its own is hardly sufficient to generate the effect—water moving in this or that way—the same is true of water on its own, taken apart from these shapes. Similarly, motion on its own, i.e., formal motion *qua* substance, produces nothing; it must be 'modified' (diversified, particularized) by the bodies that receive it.[13]

Régis's treatment of motion as a substance is one of his most important and puzzling claims.[14] The ontology of motion is among the most obscure issues in the period, and Régis deserves credit for facing the issue head-on. Régis of course recognizes that it sounds odd to call motion a substance. He responds that in so far as a created substance is one that depends only on God for its existence, formal motion, in the sense of indivisible, non-successive motion, meets this definition.[15] But motion's claim to substancehood, it seems, must end there, since it is hard to see how it could play the other fundamental role of substance, a foundation of modes.

If we read Régis's claim as merely that formal motion is in one respect *analogous* to substance, we can see how his view might make sense. Motion in the object moved is nothing but the successive impact of bodies on one another. Now, we can ask what the moving force of these collisions is; and of course, it is God. But these collisions would not take the course that they do, and the bodies would not be deflected in the directions and at the speeds they are, were it not for their modes. So the relation between what God immediately produces—formal motion *qua* substance—and the effect that actually takes place is analogous to, though not identical with, that between a substance and its modes.

I suspect, however, that there is still some sense to be made of formal motion as (literally) a substance. I have argued above that Descartes conceives of the substance–mode relation in terms of the determinable–determinate relation. On this view, a mode is, as the Latin indicates, a way in which a thing exists. At any given moment, there must be some determinate set of modes a substance possesses; but the substance itself is nothing over and above these modes. The substance *qua* determinable is an abstraction in the sense that it cannot exist without being fully determined. For Descartes, the distinction between a substance and its essence is merely conceptual. 'Thought and extension can be regarded as constituting the natures of intelligent substance and corporeal substance; they must then be considered *as nothing else but* thinking substance itself and extended substance itself—that is, as mind and body' (AT viiiᴀ. 30–1/CSM i. 215; my emphasis). By contrast, thought and extension may be thought of as modes of a substance

---

[13] *URF* 412; cf. Descartes's *Principles*, ɪɪ. 41.

[14] It is worth noting that Hobbes had charged Descartes with treating motion as a substance. In treating determination of motion as a mode of motion, Hobbes argues, Descartes is committed to this 'absurd' position. Descartes replies that 'there is nothing improper or absurd in saying that one accident is the subject of another accident' (Letter to Mersenne for Hobbes, Apr. 21, 1641, trans. Gabbey 1980: 257).   [15] *URF* 297; see Schmaltz (2003).

'in so far as one and the same mind is capable of having many different thoughts; and one and the same body, with its quantity unchanged, may be extended in many different ways' (AT viiiA. 31/CSM i. 215). To call a determinate mode of a substance a mode is simply to mark the fact that that substance may exist in many different ways; it does not imply that the substance is a bare particular or featureless substratum in which the mode must inhere.

If we suppose that Régis has this same model in mind, his treatment of formal motion as a substance becomes intelligible. Régis, on this reading, identifies formal motion *qua* substance with the essence of motion, i.e., with motion as a determinable, as opposed to any of its modified instantiations. Now, no determinable can exist except as determined in particular ways; this is precisely what the modes of created substances contribute. We then get the result, just as Régis requires, that formal motion *qua* substance is nothing; God could not produce it without also producing the modes of bodies that make it fully determinate. Nevertheless, we can isolate, in thought, the contributions to formal motion made by God and creatures by understanding it as a substance that supports modes in the way that a determinable essence supports determinates.

Note that this brand of concurrentism is quite different from the view suggested in Descartes's *Le Monde*. On that view, God creates rectilinear motion and relies on the modes of bodies to divert it in other directions. Régis would reject this, since motion in whatever direction requires a second efficient cause. There is nothing special about rectilinear motion.

Régis's God, then, produces immediately both the substance and essence of modified things ('les choses modales'); he produces the (fully modified) existence of these same things mediately, by secondary efficient causes.[16] It is not that the secondary cause literally transmits its motion to another; rather, all we can mean by this talk of 'transference' is that the body that is struck takes on a particular quantity of motion which, while numerically distinct *qua* mode from the motion of the striking body, is nevertheless quantitatively identical.[17]

It is worth pointing out that Régis rejects the cinematic model and so is faced with the problem of the transference of motion. If motion were nothing over and above the successive re-creation of beings in new locations, there would be no such problem, as I have argued; there is no need to suppose that a mode can migrate from one body to another. Régis wishes to avoid both this occasionalist model and the obscure influx model, which seems committed to scholastic real qualities. But it is not clear that he has much in the way of a positive view about the ultimate ontological status of motion and its apparent communication from one body to another. As we shall see, Glanvill and Locke will make much of this in exposing the explanatory limits of corpuscular mechanism.

---

[16] '[D]ans l'ordre de la nature, Dieu produit toutes les substances et les essences des choses modales immediatement par luy-même, et . . . il ne produit l'existence de ces mêmes choses modales que par des causes secondes' (*URF* 271).   [17] *URF* 299.

We are now in a position to work back to one of our main themes: the contrast between top-down and bottom-up accounts of the natural world. Régis is intriguing partly because he is struggling to maintain the ancient, power-driven account of the workings of material beings. Unlike Descartes and Malebranche, he wants to combine ontological and course-of-nature mechanism.

The course of nature, for Régis, is hypothetically necessary in the sense that once God creates and conserves a world with a given set of objects and modes, and produces the substance of formal motion, there is only one way in which things can happen. What happens in that world depends on the way in which motion is realized in the natural world, and this is a function both of God and of the modes of extended substances. The course of nature is not the sole consequence of God's nature or volitional acts. If God were to create a material substance that somehow lacked the mechanical properties that serve as secondary efficient causes, or simply to create and conserve material bodies with different locations and modes, the course of nature would be different. Thus—and here is the contrast with both Descartes and Malebranche—the only way for the course of nature to change is for the secondary causes to change.[18]

This is not to say that God does not perform miracles. And on this point, as on the question of mind–body interaction, Régis retreats to faith. That is, he does not think it is possible for reason to reconcile the immutability of the course of nature and the performance of miracles; being assured of the latter through revelation, however, he must accept them. Knowing by reason that the laws of nature are immutable and by revelation that God has changed a rod into a serpent, 'I am obliged to accept these two truths as consistent even though I cannot conceive how they can be reconciled.'[19] This open confession of irrationality is hardly satisfying; it serves, however, to illustrate his firm commitment to the hypothetical necessity of the course of nature.

At this point one might begin to doubt whether there is much cash value to the distinction between concurrentism and occasionalism. Malebranche, for example, takes natural causes, which themselves are impotent, to be prompts for God to perform a given action; they are thus background conditions to this action, given that God chooses to act in accordance with these occasional 'causes.' And Malebranche admits that those 'philosophers who assert that secondary causes act through their matter, figure, and motion . . . are right in a sense' (*SATE* XV: 658). Indeed, the charge that Régis's secondary causes are just occasional

---

[18] See *URF* 243, 277.

[19] 'Je seray obligé de recevoir ces deux veritez commes tres constantes, bien que je ne puissse pas concevoir comment elles s'accordent ensemble' (*SGP* 93).

causes was brought against him only four years after his *Système* appeared in print.[20] Nevertheless, I think there is a substantive difference between these views.

One way to draw out the causal contribution of creatures is to ask what God would have to do to make the course of nature different. Take two qualitatively identical, purely material worlds. Must their futures be identical? For Malebranche, the answer is no. What fixes the course of nature is only God's nature. If his nature were different, Malebranche's God could produce motion and yet choose to distribute it in a different way in each world.[21] For Régis, the futures of these worlds must be identical, even though neither of them is in itself, of course, a necessary being. That is, God need not have created either of them; but once he does, the events within those worlds must be exactly the same, so long as he creates the substance of formal motion. So whether one wishes to count second efficient causes as genuine producers of effects or as background conditions is, while important, not the crucial issue in the debate between concurrentism and occasionalism. The real question is, what fixes the course of nature? And to this Régis and Malebranche give very different answers.

[20] Henri de Lelevel writes, 'Monsieur Regis accoûtumé à discoursis sans preuve & sans fondement, établit deux sortes de causes *efficientes*. Il en veut aux *causes occasionnelles*, mais malheureusement, ou il ne sçait ce qu'il dit, en les voulant détruire, où il n'en fait que changer le nom, en les appellant causes *efficientes secondes*' (1694: 121–2).

[21] Malebranche, of course, is at pains, particularly in the *Treatise on Nature and Grace*, to emphasize that God must distribute motion in the way that he does, given his nature as omnipotent and omnibenevolent. This does not affect the main point I am making, namely, that the distribution of motion depends on God's will and nature and not on secondary causes.

14

Régis Against the Occasionalists

We can fill in Régis's positive view by seeing how it fares in light of its main competitors in the context of Cartesianism. Let us begin with Régis's central arguments against his opponents:

1. At least in the case of Descartes, it is difficult to see how the manifest diversity of natural effects can be reconciled with God's immutability. Consider again Descartes's claim 'From God's immutability we can also know certain laws or rules of nature, which are the secondary and particular causes of the various motions we see in particular bodies' (AT viiiA. 62/CSM i. 240).

Régis would regard this as self-contradictory. '[A]ll that is immutable in the production of effects must be attributed to God as the first cause, whereas all that is changeable must be attributed to body as the second cause.'[1] The varied motions of bodies cannot be attributed to laws of nature in Descartes's sense, for these follow immediately from God's nature alone, and their immediate effects or consequences must likewise be immutable. Now, Descartes claims in the same passage from the *Principles* that 'God imparted various motions to the parts of matter when he first created them.' The full story of the variety of motion that we observe must invoke, then, not only God's willing the laws of nature and preservation of moving bodies, but also his creation of a diversity of motions in the initial state of the universe. It is the last of these that Régis regards as unintelligible. An explanation of that which is successive and changeable in bodies requires a principle equally successive and changeable.

Aquinas had used roughly the same consideration in arguing against the view that created beings are bereft of causal power. '[I]f God works alone in all things, then, since God is not changed through working in various things, no diversity will follow among the effects through the diversity of the things in which God works' (*SCG* 69: 125).

Ultimately, as I have argued above, any view that posits an immutable God will run up against some version of this problem. It is hard to see how to account for the variety among God's effects whether one is a concurrentist or an

---

[1] '[T]out ce qu'il y a d'immuable dans la production des effets doit estre attribué à Dieu comme à la cause premiere, au lieu que tout ce qu'il y a de changeant, doit estre attribué aux corps, comme à la cause seconde' (*URF* 298).

occasionalist. Arguably, however, this problem is less pressing for Régis than it is for Descartes, since Régis can at least appeal to the diversity of modes to account for the diversity of effects. But unless matter is co-eternal with God, something none of these thinkers would have accepted, the problem remains.

The next arguments are directed chiefly at Malebranche:

2. Occasionalism makes the merely apparent secondary causes, which Malebranche termed 'natural causes,' otiose. It would be silly, Régis argues, to suppose that fire and other natural agents have 'this innumerable diversity of qualities, these powers so different and yet at the same time so well proportioned to their effects'[2] only to serve as occasions for the sole cause to act.[3] If there is a diversity of natural, material things and their properties, the simplest explanation for the course of nature adverts at least partly to these things. And of course God, being perfect, always acts in the simplest ways.

Arguably, the best response to this kind of objection comes not from Malebranche but from Berkeley. Faced with the worry that, given the denial of causal powers to bodies, 'we must think all that is fine and artificial in the works, whether of man or nature, to be made in vain' (*PHK* 1. 60), Berkeley appeals to the need for regularity in nature. Although God need not have created the exceedingly intricate structures we observe in nature, given his omnipotence, he nevertheless chooses to do so to preserve the constancy and regularity that alone makes experience intelligible to us (*PHK* 1. 62).[4] Régis and Aquinas would find this simply implausible; although God's regular action might require some degree of complexity in the structures on which he acts, it is difficult to believe that it requires *this much* complexity.[5]

3. Occasionalism supposes that natural events can be caused by God immediately, and without the cooperation of the powers of bodies. But the modes of matter that play the role of secondary causes are indispensable. It is 'an incontestable truth' that bodies 'cannot act without certain dispositions.'[6] On its face, this argument seems weak, at best; the occasionalist can of course grant Régis's

---

[2] 'cette diversité innombrable de qualitez, ces virtus si differentes, et en même temps si proportionnées à leurs effets' (*URF* 411).

[3] This argument is drawn directly from Aquinas; see *SCG* 69: 125 ff.: 'It is contrary to the notion of wisdom that anything should be done in vain in the works of a wise man. But if creatures did nothing at all towards the production of their effects, and God alone wrought everything immediately, other things would be employed by Him in vain for the production of these effects.'

[4] Although it seems odd to invoke Berkeley here, since he regards material beings as otiose, the form of the problem he confronts is precisely parallel to that faced by Malebranche: couldn't an omnipotent being have come up with a more straightforward system than what we observe in nature, given that ordinary objects have no causal power? Whether 'ordinary objects' here refers to extended beings or to ideas, the problem is the same.

[5] One possible line of resistance is Leibnizian: one might claim that the best possible world is the one with the most variety and diversity.

[6] 'une verité incontestable', 'ne peuvent agir s'ils n'ont de certaines dispositions' (*URF* 411). Here, I take 'disposition' to mean the arrangement of the bodies' parts.

conditional claim—if bodies act, it is in virtue of their modes—and deny that the antecedent is ever satisfied. But all Régis needs to strengthen his argument is the claim that God acts on bodies only through motion. 'For motion considered in itself produces nothing; it needs to be modified to be efficacious. What modifies the motion is thus a very real and positive physical cause.'[7] Substantial formal motion on its own can accomplish nothing. As we shall see below, Régis accepts a key premise of Malebranche's 'no necessary connection argument,' namely, the claim that there is a logically necessary connection between a true cause and its effect such that the denial of a proposition stating this causal relation is a contradiction. The true cause, for Régis, will be God in his capacity as the source of motion considered in itself, created substances, and their modes, which are responsible for the way in which this motion itself is modified. These are individually insufficient to bring about any physical event: it is inconceivable that God create motion in bodies without the cooperation of secondary causes.

4. The final argument I shall examine is directed specifically at Malebranche. Like Malebranche's own divine concursus argument, it concerns the contents of God's volitions. Does God will the existence of each and every thing individually? Or does he simply will the laws of nature, taken as conditional claims, plus the continual re-creation of bodies? Régis poses a dilemma. God's volitions will either be particular (where this means that God wills a particular, fully spatiotemporally indexed object or event as such) or general (where God wills only the laws of nature plus the existence of all beings). Neither is consistent with God's nature. To take each case in turn:

(*a*) In the context of particular volitions, occasional causes are incompatible with God's immutability. If an occasional cause determines God to act in ways that, absent that cause, he would not, this 'supposes in God an indetermination that is incompatible with his immutability.'[8] That is, if God were to will a particular event in response to an occasional cause, the content of his will would depend on something outside of him, and this is impossible. A second consideration Régis advances is that this multiplicity of volitions conflicts with God's simplicity.[9] This is hardly fair to Malebranche, for there is no reason why God must have a number of distinct volitions, each directed at distinct objects or events, as opposed to a single super-volition whose content referred to all of these.

(*b*) Whereas particular volitions conflict with God's simplicity and immutability, general volitions conflict with God's actuality. 'These general volitions would be of themselves indeterminate and this contradicts the simplicity and actuality

---

[7] 'Or le mouvement consideré en luy-même ne produit rien; il a besoin d'estre modifié pour estre efficace. Ce qui modifie le mouvement est donc une cause physique tres réelle et tres positive' (*URF* 412).

[8] 'suppose en Dieu une indetermination qui est incompatible avec son immutabilité' (*SGP* 110).

[9] *URF* 92.

of the divine nature.'[10] When God forms a general volition, its content has in no way been determined by any particular agent, and so there is no conflict with God's immutability. God simply wills, e.g., 'All motion will continue in a straight line unless the moved body is interfered with.' A general volition would thus be a volition whose content was not fully specified. But then God's will is not fully determinate and thus not fully actual.

Arguments 2 and 3 seem to me to be decisive. (I leave it to the reader to work out how Malebranche would reply to (4), given his two conceptions of laws.) Of course, it is hardly news that occasionalism faces grave philosophical problems, even if God's existence were taken for granted. But Régis's arguments point to genuine defects in the Cartesian approach whose amelioration requires something very like Régis's own concurrentism.

However attractive Régis's view is, at least within its historical context, it is not obvious that it can withstand the occasionalist arguments. The chief worry here is that Régis is committed to enough of the ontology and metaphysics of causation deployed in Malebranche's arguments to make a slide into the wholesale denial of causal powers to finite beings inevitable. Whether and how Régis is able to resist this is not only of intrinsic interest but has implications for those philosophers, like Locke, who also wish to invert the Cartesian top-down picture. Let us begin by considering the divine concursus argument before turning to the 'no necessary connection' argument.

The former argument, as we have seen, moves from God's role as sustainer of the universe to the conclusion that only he is a cause. For in re-creating beings, God must re-create them in determinate places; and if this is so, there seems to be no way for him to cooperate with creatures. I have argued above that, while Descartes resists full-blown occasionalism, his view does not have the resources to respond to this argument. Régis, however, can do somewhat better.

To reconstruct Régis's response, we must begin with premise 9 of that argument: To conserve a being is to re-create that being. Only in this way can Malebranche derive (10): since God must re-create the world anew at each moment, he must choose determinate places for every being he conserves. Régis's strategy will be to drive a wedge between creation and conservation in order to bring out the corresponding difference between God's immediate and mediate actions.

For Régis, creation is an act whereby God produces something immediately;[11] in this sense, creation concerns only the existence of substances, not their modes (or better, substances considered in themselves and not as 'choses modales'). Conservation, by contrast, concerns what God produces mediately, i.e., modified

---

[10] 'Ces volontez generales seroient de soy indeterminées; ce qui repugne à la simplicité et actualité de la nature divine' (*SGP* 92).   [11] *URF* 952.

things. Conservation results, not in the existence of substances considered absolutely, but in 'modes that diversify the substance by motion, and that give it new forms, such as those of stone, wood, etc.'[12] Régis is not, however, saying that there is a real distinction between a substance and its modes, as if they were independent beings; this would just be the mistake of real qualities again. A thing's modes are 'nothing but the subject or substance itself.'[13] God need not immediately will the substance's modes; what he immediately creates is the substance itself, and its modes will be determined by the ways in which the second causes modify the motion that God produces. But this is not to say that he does not also will the substance's modes. He does, and he must, by the determinateness principle (premise 4), which I think Régis would accept. The key is that God *mediately* wills these further effects, in the sense that he uses created beings as instruments to achieve his ends. There is thus no indeterminacy in the content of God's volitions, and hence no sin against the conceivability or determinateness principles. Such a reply is not open to Descartes, since he simply does not have this concurrentist machinery.

One might worry that in distinguishing between God's mediate and immediate actions, Régis departs from concurrentism and retreats to conservationism, which holds that God merely conserves bodies while their powers operate autonomously. On this view, God's activity is demoted to a mere necessary condition, and the notion that God 'works through' bodies rather than merely permitting them to pursue their own directions is abandoned. Scholastic concurrentists like Suárez often cast the debate in terms of immediacy, taking their opponents as claiming that God acts only mediately in ordinary events.[14] I think the difference between Régis and Suárez here is merely verbal. Régis's God is not just another necessary condition; he is genuinely active in the production of effects. If anything, Régis's mechanistic view makes even more stark the difference between conservation and concurrence. To say that God produces modes 'mediately' is to say that he produces them through the activity of secondary causes, not that he merely allows those secondary causes to persist.

We can now turn to Régis's response to the 'no necessary connection' argument. Malebranche's cognitive model, which holds that cause and effect must be linked by an identity of intentional content and the state of affairs it refers to, drives NNC. Only in this way, he thinks, can the *liaison nécessaire* be preserved in the context of mechanism. I suggested above that a promising initial reply to NNC would try to deny the conceivability of the states of affairs that seems to sever this necessary connection. This, I believe, is how the traditional concurrentist (that is, one who retains Aristotelian powers) ought to reply. Our

---

[12] 'modes qui diversifient la substance par le mouvement, et qui luy donnent des nouvelles formes, telles que sont celles de la pierre, du bois, etc.' (*URF* 334); see 322–3 for the response to Descartes.

[13] 'n'est autre chose que la sujet, ou la substance même' (*URF* 960).

[14] See Suárez, *Metaphysical Disputations*, 22. 1, in Suárez (2002).

question, then, is whether Régis can mount such a reply *without* such powers, or with a suitably reductive account of them. Another way to put this question is to ask whether Régis's view has the resources to supplant the cognitive model and offer a perspicuous account of the *liaison nécessaire* without implying that bodies are possessed of intentional states.

Régis's task, then, is to resist Malebranche's argument without appealing to any materials beyond those available to a mechanist. He attempts this, I believe, by invoking the geometrical model of causation. Régis writes,

> We should not say that we see no necessary connection between the second causes and the effects we attribute to them, such as we see between the first cause and its effects. For unless we renounce the senses and reason, we must admit that we see an obvious connection. We see, for example, that the production of flour is necessarily connected with the way in which the mill changes the motion of the water and wind that comes immediately from God. We see, again, that a house one builds is necessarily connected both with the way in which the motion of the stones is modified and with this same motion, and so on for all the other effects God produces by the second causes, as by instrumental causes.[15]

While this has an air of table-thumping about it, one can see Régis's point: there is something obvious about the claim that the direction of motion, itself derived from instrumental causes, is necessarily connected with the effects we observe. Note that the necessary connection is not said to hold between a given effect and its total cause, which would include both God and modal things. Rather, there are two distinct causal orders, each characterized by necessity: primary cause–primary effect (God–the substance of formal motion) and secondary cause–secondary effect (modes of created beings–modification of formal motion). Malebranche and Régis agree with regard to the primary causal sequence; God, being omnipotent, cannot fail to produce anything that he wills. But what of secondary causes?

By now, it should not be surprising that Régis does not take the route that naturally occurs to us, namely, to deny that genuine causation involves logical necessity. Régis might have preserved his claim that there is a necessary connection between cause and effect by drawing a distinction between logical and nomological necessity. Whether or not such a view is available to him (and I see no reason to think it is), the concurrentist tradition in which he is steeped has no room for non-logical necessity where causation is concerned. Instead, Régis

[15] 'Qu'on ne dise donc pas qu'on ne voit point de liaison necessaire entre les causes secondes et les effets qu'on leur attribuë, comme l'on en voit entre la cause premiere et ses effets; car à moins de renoncer aux sense et à la raison, on y voit une manifeste. On voit, par example, que la production de la farine est aussi necessairement liée avec la maniere dont le moulin modifie le mouvement de l'eau et du vent qui vient immediatement de Dieu. On voit encore qu'une maison qu'on bâtit, est liée aussi necessairement avec la maniere dont se modifie le mouvement des pierres, qu'elle est liée avec ce même mouvement; et ainsi de tous les autres effets que Dieu produit par les causes secondes, comme par des causes instrumentales' (*URF* 414–15).

insists that the necessary connection, in the strongest sense, is in fact present, even between secondary causes and their effects. To this extent, Régis, like Malebranche, retains a key feature of scholasticism. The necessary connection, however, holds true not because of any connection between Aristotelian powers or volitions and their effects, but only in virtue of the mechanical properties of bodies. Can Régis then offer his own version of the Aristotelian response, namely, that when conceiving that these connections fail to hold, Malebranche is misdescribing what he has conceived?

For Régis's response to work, in no possible state of affairs with the same arrangement of secondary causes can any effect but one be generated. Now, secondary causes do nothing on their own, so it is easy to imagine a world with the same arrangement of modes and yet no effects whatsoever, or effects only in certain regions of that world, since God might simply fail to produce the substance of formal motion. Precisely because the secondary causes are dependent on the first cause, there cannot be a necessary connection between the elements of the secondary series. But perhaps this isn't what Régis means. In his examples above, he always includes the primary cause and its effect. So the real question is whether there can be two (mechanically) identical worlds, with the same amount of motion, whose events nevertheless proceed in distinct ways. Are there conceivable, and hence possible, worlds in which the modes of matter are held constant, as is the production of motion per se, and yet the course of events is different?

We must have a bit more of Régis's view on the table before we can answer this question. Although Régis will have no truck with occult qualities, he does maintain a functional replacement for scholastic forms.[16] Forms will still be that which makes a thing the kind of thing it is and explain why it has the properties it does.[17] But instead of existing as something over and above matter, a form is just a mode of that matter and hence 'nothing but the subject or the substance itself.'[18] Forms are ontologically innocuous, then, because they can be reduced to elements any mechanist already accepts.

How is this reduction accomplished? Régis argues that what makes gold gold, for example, is 'a certain order and arrangement of parts'[19] that gives it properties it would otherwise not have.[20] But there is nothing mysterious in this: to say that gold has the (passive) power to dissolve in aqua regia is just to issue a promissory note for an explanation in microphysical terms that would appeal to nothing

---

[16] Régis does deny substantial forms, but by this he means forms that are themselves substances, i.e., real qualities. Thus, depending on how one reads Aristotle on the issue, Régis is not being disingenuous when he says, 'we say nothing that contradicts the views of Aristotle' ('nous ne dirons rien de contraire au sentiment d'Aristote'; *SGP* 393).   [17] *SGP* 391–3.

[18] 'n'est autre chose que le sujet, ou le substance ce même' (*URF* 960).

[19] 'un certain ordre & arrangement de parties' (*SGP* 392).

[20] The seeds of this sort of view are present in both Descartes and his followers and made them a target of abuse for Glanvill and Newton, among others. See above, sect. 3.3.

but the size, shape, and movement of the relevant bits of matter. And there is nothing mysterious either in a body's gaining or losing one of these forms, for 'all generation and corruption that happens in the world can be explained by local motion alone.'[21] Unlike Aristotelian essences, such forms are not migration barriers. On the Aristotelian view, possession of a given essence prevents a subject from acquiring a new such essence without being destroyed. Régis thinks such migration is a matter of the rearrangement of microstructural parts; in principle, nothing prevents lead from being transmuted into gold. Malebranche himself accepts forms so understood.[22] We can think of these ersatz forms as second-order modes, that is, properties accruing to bodies in virtue of their first-order modes.

Régis thinks it is impossible for us to know these forms, to know what microstructural arrangement gives gold its distinctive properties, simply because the parts in question are insensible.[23] (We can, of course, make reasonable conjectures about these structures.) With this on the table, we can see how Régis would reply to Malebranche's elimination argument. A thing's powers are identified with its modes; this makes sense partly because those powers are construed on the concurrentist model (and so the inactivity of bodies left to themselves is preserved) and partly because Régis simply rejects Malebranche's account of modes as relations of distance. The argument from nonsense will also have been answered, since Régis can in fact give a perspicuous account of powers. A thing's powers will consist in its modes (and those, presumably, of the bodies around it); like forms, powers are second-order properties, features that accrue to a body in virtue of its (first-order) modes.

How does any of this help with NNC? The challenge is to explain away the apparently conceivable, and hence possible, states of affairs that threaten the necessary connection between cause and effect. With his account of forms and powers in hand, Régis can account for the appearance of conceivability presented by counterfactual statements in purely epistemic terms.[24] Our ignorance of the underlying mechanical disposition of a hunk of gold means that we can

---

[21] 'toutes les generations & corruptions qui arrivent dans le monde se peuvent expliquer facilement par le seul movement local' (*SGP* 393). Régis continues: 'or, put better, solely by the transposition of imperceptible parts of matter, by which they are diversely figured and arranged, making them admit of different properties' ('ou, pour mieux dire, par la seule transposition des parties imperceptibles de la matiere, lesquelles selon qu'elles sont diversement figurées & arrangées, rendent leur sujet capable de differences proprietez').

[22] 'I should point out here in passing that there is nothing wrong with the terms *form* and *essential difference*. Honey is undoubtedly honey through its form, and in this lies its essential difference from salt. But this form or essential difference is only a matter of the different configurations of its parts . . . and although matter in general has the configuration of the parts of honey or of salt, and hence the form of honey or of salt, only accidentally, yet it can be said that a given configuration of their parts is essential for honey or salt to be what they are' (*SAT* I. xvi. 4: 75).

[23] *SGP* 392.

[24] There is a complication here I am ignoring. In responding to conceivability arguments, one always has the choice of either denying that the state of affairs in question is in fact conceived (which is how I am putting it), or denying that the kind of conceivability in question entails or justifies possibility. I don't see much difference between these. If the reader prefers, she is free to read me as arguing that conceivability gets us only epistemic possibility.

seem to conceive of a state of affairs in which a hunk of matter with precisely the same microstructure is not dissolved in aqua regia. But this appearance of conceivability does not entail possibility, any more than my (at least apparent) ability to think of the claim that Hesperus is not Phosphorus entails that there is a possible world in which they are not identical.

So although Régis replaces scholastic forms with mechanically acceptable substitutes, he nevertheless holds that when a given ersatz form is realized in a bit of matter, there is no logically possible world in which that body behaves in any way other than it does in the actual world. Worlds in which gold does not dissolve in aqua regia, fire fails to burn, and so on, cannot really be conceived, though they can be described in a superficial way. For fire that failed to burn would not be fire at all, i.e., it would not have the particular modes at the microscopic level that make fire what it is, and make it do what it does.

One might still feel, however, that Régis has missed the point of Malebranche's argument. For the larger point, embodied in both NNC and the epistemic argument, is that finite beings are not of the right ontological type to serve as causes, since they lack the structure and content of intentional states. Here I think Régis can do no more than dig in his heels against the cognitive model. Not all connections between events are secured by an identity of content and represented state of affairs; causal connections, in the bodily case, are secured by the structure of the bodies themselves. What would really be needed here, it seems to me, is a thoroughgoing reduction of powers that explains, or explains away, their apparent *esse-ad*. That is, to do more than stamp one's feet and insist on the logically necessary connection between mechanical modes, one would have to explain just how and why that connection obtains. Such a project would require an analysis of relations generally that purges them of the ontological spookiness Malebranche and Descartes point to in the 'little souls' argument.

As we shall see, Locke presents a much more developed picture of the kind of necessity present in the natural world. Carrying out this bottom-up project requires a clear-eyed statement of the geometrical model and an ontology of relations suitable to mechanism, both of which are merely implicit in Régis.

Faced with the host of problems that define the Cartesian predicament, Régis returns to what otherwise seems a moribund scholastic tradition, that of con-currentism. Indeed, it would be difficult to find a defender of any recognizable version of concurrentism after Régis.[25] Thus, from one point of view Régis seems hopelessly backward, trying to cling to the scholastic raft long after the attacks of Descartes, Malebranche, and others have shot it to pieces.

But from another point of view, Régis can be seen as trying to preserve what was valuable in the Aristotelian tradition. By 'sanitizing' key concepts such as power and form and making them intelligible in the context of mechanism, Régis

---

[25] That is, after 1704, the publication of *URF*. Leibniz is a possible exception here.

preserves something like a commonsense view of the powers of physical objects. The 'bottom-up' conception of the workings of the natural world is independently plausible, and makes room for scientific explanations of phenomena in terms of the modes of the objects involved.

Much of Régis's work is in effect a promissory note for a fuller working out of the reduction of powers. Boyle and Locke, I shall argue, attempt to make good on this promise. But to do so they have to contend with a host of further attacks, which seem to show up mechanism as a hopelessly incomplete account of the natural world. It is to these challenges, and the others we have already set out for the bottom-up view, that we must turn next.

PART III
POWER AND NECESSITY

INTRODUCTION

With this part of the book, we leave behind the Cartesian debate as it was conducted on the Continent. Many of the same themes were to be sounded in Britain, but there the debate has a decidedly different cast. For the British environment saw increasing dissatisfaction with Cartesianism broadly construed, and as a result, some novel views were invented.

A chief source of discontent in this context is, of course, the patina of Epicurean atheism Cartesianism came to acquire. To some degree, this is no doubt due to simple misreadings, willful and otherwise, of Descartes. Much more important, however, is the influence of Hobbes.[1] In the so-called *Short Tract* and parts of *De Corpore*, among others, Hobbes presents a thoroughgoing materialism cast in the mechanist mold. Like Régis, Hobbes adapts Aristotelian notions to his own ends, giving a reductive account of power in terms of the modes of created beings. But unlike either Régis or the main figures of this part of the book—Boyle and Locke—he does not present a very detailed picture of the causal powers of bodies.[2] Although Hobbes will have cameos in what is to come, our present purposes are best served by looking at the reactions his view inspired.

Figures such as Cudworth, More, Glanvill, Boyle, and Locke all, to one degree or another, deplore Hobbes's rejection of an immaterial being, not just for religious reasons, but because they find the world as it is inexplicable in purely mechanical terms. Even if, as I shall argue, Locke defends a broadly deist conception of the working of God, he nevertheless argues that God himself must be an immaterial being; otherwise, many features of the world, particularly the presence of cognition, would be inexplicable (*Essay*, IV. x). And even if matter in motion has some powers of its own, these would not suffice to explain 'the admirable conspiring of the several parts of the universe,' as Boyle puts it (RD 157).

But this is in many ways the least interesting aspect of the revolt against Hobbesian mechanism. As we shall see in the next chapter, many philosophers of the time argue that corpuscularianism is subject to insuperable difficulties as a hypothesis in its own right. Cudworth, for example, argues that 'there are many . . . particular phenomena in nature, as do plainly transcend the powers of mechanism, of which, therefore, no sufficient mechanical reasons can be devised.'[3] Even the cohesion of the parts of bodies turns out to be a serious problem.

This is not to say, however, that Cudworth and More favor anything like Descartes's own position as an alternative. If religious motives were all, this would be surprising, since Descartes's body–body occasionalism, as well as his

---

[1] On this, see esp. Jesseph (2005).   [2] Some would disagree; see esp. Leijenhorst (2002).
[3] Cudworth (1837: i. 210).

top-down picture of the laws of nature, could hardly incorporate God more fully. Neither Cudworth nor his contemporaries, with the exception of John Norris, finds occasionalism at all palatable. Nor do they show signs of returning the scholastic concurrentism. Having rejected Hobbism as well, what options are left open to them?

Recall that Cudworth is the first figure to pose the argument from the laws of nature. There, he argues that Descartes must either suppose that the laws of nature execute themselves or resort to occasionalism, each of which is manifestly absurd. Now, if these laws are neither features of God's will nor autonomous beings governing events, they must be put into action by a further principle or force. Cudworth invents his 'plastick nature,' a sort of 'deputy God,'[4] to put these laws into action. 'Nature is art as it were incorporated and embodied in matter, which doth not act upon it from without mechanically, but from within vitally and magically.'[5] For his part, More introduces his 'hylarchic principle,' a source of action that is neither mechanical nor divine, but somewhere in between.

The details of their views need not detain us. What is interesting for our purposes is the response they offer to what I have been calling the Cartesian predicament. In their way, both seek to offer a bottom-up picture, which makes the properties of objects themselves, and not the divine will, the immediate source of their behavior. Thus, they preserve course-of-nature mechanism—what happens, happens because bodies are as they are—at the cost of ontological mechanism. Just the reverse was the case with Descartes and Malebranche; for them, ontological mechanism was non-negotiable, and issued in the denial of course-of-nature mechanism.

If Descartes had nothing but scorn for the 'little souls' of the scholastics, it is not hard to imagine what he would have made of 'plastick natures' and 'hylarchic principles.' More and Cudworth make their escape from the Cartesian predicament only by peopling nature with principles neither mechanical nor divine, a view straightforwardly at odds with the ontological mechanism our previous philosophers were striving so hard to preserve. It is not surprising, then, that Cudworth and More draw the same accusations of obscurantism that dogged the Aristotelians. As Glanvill puts it, mocking Cudworth's innatist epistemology: 'The *Plastick* faculty is a fine word, and will do well in the mouth of a puzzled *Empirick*: But what it is, and how it works, and whose it is, we cannot learn; no, not by a return into the *Womb*.'[6]

The particular problems we shall explore, then, are powerful motivations for the rejection of ontological mechanism. Coupled with the uncertain role of God in anything that smacks of Hobbism, they form the most immediate stumbling

---

[4] The phrase is Jesseph's (2005).   [5] Cudworth (1837: i. 220).

[6] Glanvill (1665: 33). '*Empirick*' here probably means a doctor who prescribes remedies without the slightest idea how they work; cf. Locke, *Essay*, IV. xx. 4: 709.

blocks for anyone trying to carry out a thoroughgoing mechanist program. Striking a blow from yet a different angle, Cudworth's argument from the laws of nature challenges any ontological mechanist to hang on to course-of-nature mechanism, for absent a 'plastick nature' or God's direct involvement, it is hard to see how these laws can be enforced.

Thus, even aside from religious questions, mechanism in Britain is in fairly dire straits by the time Boyle and Locke come along. Their solution, or more properly Locke's, I shall argue, is to reconcile ontological and course-of-nature mechanism by means of a reductive view of relations generally, and powers in particular. The laws of nature are nothing but generalizations over the powers of bodies and so need no enforcer to put in them into action. If successful, Locke will have negotiated a path through the modern thicket and arrived at a wholly novel view: a thoroughgoing bottom-up mechanism that is neither concurrentist nor 'magical.'

15

'A Dead Cadaverous Thing'

Cudworth's attack on mechanism is not limited to its Epicurean variety. Even mechanical theists, he believes, have 'an undiscerned tang of the Mechanic Atheism' about them, 'in that their so confident rejecting of all final and intending causality in nature, and admitting of no other causes of things, as philosophical, save the material and mechanical only; this being really to banish all mental, and consequently Divine causality, quite out of the world; and to make the whole world nothing else, but a mere heap of dust fortuitously agitated, or a dead cadaverous thing, that hath no signatures of mind or understanding.'[1]

'Final and intending causality' includes not just the kind of goal-directedness Cudworth thinks is so obvious in the natural world but causality operating with a degree of intentionality or representational power. Cudworth's growing disenchantment with Descartes was in part due to the latter's explicit rejection of final causality as a useful tool in natural philosophy. In response, Cudworth introduces his 'plastick nature,' which, despite not being a full-fledged mind, is a kind of intermediary between God and the world. It is as if Cudworth had taken up Malebranche's challenge to 'make a man out of his armchair' and clamped down, with a grin, on the bullet.[2]

If Cudworth's positive story is at the outer reaches of the plausible, some of the arguments he mounts in its defense are not. In this chapter, we shall be concerned with the whole spectrum of difficulties with mechanism raised, not just by him, but by More, Glanvill, and Locke himself. First among these is the problem of cohesion: how, exactly, does the mechanist propose to explain why certain bits of matter stick together more strongly to some than others? Malebranche, though not a course-of-nature mechanist, had made heavy weather of this topic already.[3] Second, we shall look at gravity, which is the undoing of any view that tries to account for natural phenomena purely in terms of direct impulse. Third, the connection between the primary qualities of bodies and the phenomenal states they give rise to in us is no less mysterious on the mechanical view. There are also intriguing problems about the ontology of motion, pressed by Locke. This is what I have called above the problem of transference: if motion, *qua* mode,

---

[1] Cudworth (1837: i. 209).
[2] This is not to say that the 'plastick nature' has self-consciousness, or is fully endowed with the powers of a human mind. See Cudworth (1837: i. 228 ff.).   [3] *SAT* vi. ii. 9.

cannot literally be transferred from one body to another, what is going on when two bodies collide? Finally, we shall have to mention the objection from the supernatural, of which More makes much.

Two main responses are available to these problems, assuming occasionalism is off the table. First, one may, like Cudworth and More, take these problems to show the falsity of ontological mechanism and retreat to hylarchic principles (or indeed some version of occasionalism). Second, one may instead chalk these problems up to epistemic limitations and refuse to draw metaphysical conclusions from them. This is the attitude of Glanvill, who regards corpuscularian mechanism as 'an *Hypothesis*, well worthy [of] rational belief,'[4] even though, like all of its competitors, it fails to solve all of the problems that might be raised against it. As we shall see, the question whether Locke endorses a 'pure' mechanism—a combination of both ontological and course-of-nature mechanism—boils down to a choice between these two options.

Let us begin with cohesion. Among the many features of the natural world of which we are ignorant, Glanvill singles out this one for special attention; many solutions had been proposed but none found. 'If we seek for an account how the parts of these [bodies] do cohere, we shall find ourselves lost in the enquiry.'[5]

Why does gold resist the pressure to break more than, say, balsa wood? The Epicurean answer had appealed to the shapes of their micro-level constituents, which, being formed into hooks and eyes, among other figures, latch onto one another. The particles of gold contain more such bonds than do those of balsa wood. Malebranche shows that this is no help. For it merely pushes the question back, and we find ourselves confronted again with the question of cohesion at the level of these microstructural figures themselves. One might insist that they are indivisible; but this would make it not just difficult but impossible to break a bar of gold, or even a section of balsa wood, in so far as either of them contains any of these bonds.[6]

Malebranche's solution is to posit an invisible horde of 'tiny bodies,' an 'ether' whose motion compresses bodies and resists our efforts to divide the macro-level body.[7] But, as Locke argues, this proposal, like the Epicureans', simply pushes our problem back: what accounts for the unity of the parts of the ether itself? Since they by definition lack the cause of cohesion that is invoked to explain that of macro-level bodies, we are once again left with a seemingly insoluble problem (*Essay*, II. xxiii. 23: 308).

Two contiguous slabs of polished marble cannot be separated by pulling the one away from the other but only by lateral motion. Locke uses this evidence, mentioned by Malebranche himself,[8] against him (II. xxiii. 24: 309). It is just

---

    [4] Glanvill (1665: 65).   [5] Glanvill (1665: 36).   [6] See *SAT* VI. ii. 9: 513.
    [7] *SAT* VI. ii. 9: 521. Newton also invokes the ether to explain cohesion; see Gabbey (2002: 340).
    [8] *SAT* VI. ii. 9: 510. Locke might have a broader target in mind than just Malebranche; Jacob Bernoulli is another plausible candidate.

not clear why the motion of the particles of the ether should exert itself more in one direction than the other. Throwing up his hands, Locke declares that the cohesion of matter is 'as incomprehensible, as the manner of Thinking, and how it is performed' (II. xxiii. 24: 309).

Still more mysterious is gravity, whose 'cause and nature' Boyle declares 'as obscure as those of almost any phenomenon it can be brought to explicate' (RD 156). Locke, first insisting that direct contact between bodies is the only way they can interact, gradually and grudgingly modifies his view to accommodate that of 'the incomparable Mr. Newton.'[9]

The problem of gravity was well known long before Newton came on the scene. More, for instance, takes gravity to be the primary counter-example to mechanism. '[T]he *Phaenomenon* of *Gravity* is quite cross and contrary to the very first *Mechanick laws of Motion* . . .'[10] Action at a distance is anathema to any approach that seeks to analyze nature in terms of matter and motion. And Leibniz was hardly alone in insisting that Newton's conception of it represents a return to the occult qualities of the Aristotelians and a sad retreat from the perspicuity natural philosophy had in other realms attained.[11]

These are particular empirical problems that do not admit of easy explanation, or perhaps any at all, within the confines of ontological mechanism. The other two chief sources of difficulty are in a different class. First, we have the problem of the connection between primary and secondary qualities, or, minus the jargon, the now familiar problem of explaining just how and why such and such an arrangement of particles in bodies (or brains) comes to produce in us all and only the qualia that it does. As I argue below, however, this is in the first instance a problem in philosophy of mind, not natural philosophy. And it was exacerbated by the dualist presuppositions of nearly all philosophers of the age. This is quite clear in Glanvill, who diagnoses the problem in this way:

body cannot act on any thing but by *motion*; motion cannot be received but by *quantity* and *matter*; the *Soul* is a stranger to such gross *Substantiality*, and ownes nothing of these, but that is cloathed by our deceived *phancies*; and therefore how can we conceive it subject to *material impressions*?[12]

The problem of the connection between primary and secondary qualities is just a special case of the general problem of sensation that arises on any ontology of the mind but is arguably particularly acute on a dualist one. To solve it, one would need to know as much about the mind as about anything going on in the world outside the body. So although this problem is an important one in the era,

---

[9] As we shall see, matters are more complex, since the revisions he makes to his own view are actually at odds with Newton's. See below, Ch. 21.

[10] More (1662: i. 46). For more on Newton's departures from contact mechanism, see Smith (2002: 150 ff.).

[11] Alongside gravity, we might also mention magnetism; see Boyle, RD 157.

[12] Glanvill (1665: 21; see 65).

and one of which Locke in particular makes much, I see it as falling out of, or at most half *in*, the purview of natural philosophy.

Our next problem deals squarely with events in the extra-mental world. But like the previous problem, it is not straightforwardly an empirical one. This is the problem, already discussed in connection with occasionalism, of the transference of motion. When body *a* strikes body *b* and sets it in motion, is the motion acquired by *b* numerically identical with that of *a* or not? Régis, as we have seen, answers this in the negative, but he has no real account of what is going on beyond the stipulation that *b* acquires a numerically distinct mode. Above, we saw that its ability to solve this problem is one of the chief advantages of occasionalism. And even Boyle, in his mock-defense of occasionalism, makes use of the problem of transference. Not only is an appeal to other causes than God superfluous, Boyle writes, it might be unintelligible, since

it does not manifestly appear to us, that one body does really and truly move another, but only that upon a moved body's hitting another, there follows a motion in the body that is [shocked], or hitt against, nay . . . it may well be question'd, whither we can so much as *conceive* how a body can communicate motion to another and lose it, it self.[13]

As Boyle's last clause indicates, this is a conceptual, not an empirical problem. Further knowledge of the minute structures of the physical world would still leave this one unsolved. When Locke brings up the problem, it is in the context of an attempt to show that we in fact understand much less of the physical world than we think we do (II. xxiii). This is part of his larger campaign against dogmatic materialism: if the mind and its activities are incomprehensible, the natural world itself is no less so.[14] But it is tempting to take the problem to be a mere pseudo-problem. This, in fact, is Berkeley's approach: 'it is not worth disputing whether the acquired motion is numerically the same as the motion lost; the discussion would lead into metaphysical and even verbal minutiae about identity' (*DM* 68).

Finally, while it seems not just odd but comical to us, one of the main reasons some British philosophers of the age doubt the truth of mechanism is its inability to account for the supernatural.[15] Even Glanvill mentions 'Secret Conveyance,' that is, telepathy, as an instance of natural phenomena that surpass our powers of explanation.[16] More recounts the story of one Johannes Cuntius, who, kicked in the groin by his horse and subsequently expiring, became an incubus who troubled the locale female population. One lady is said to have refused his advances, saying 'thou seest how old, wrinckled and deformed I am,

---

[13] *Royal Society Boyle Papers* (Bethesda, Md.: University Publications of America), vol. 10, fos 38–40, in Anstey (2000: 211).

[14] This strategy is also in evidence in Locke's *Some Thoughts Concerning Education* (Locke 1823: ix. 184 ff.), discussed below, Ch. 21.   [15] See esp. Jesseph (2005).

[16] Glanvill (1665: 146 ff.).

and how unfit for those kinds of sports,' and Cuntius duly retreated.[17] And even Boyle did not hesitate to endorse and have printed a translation of Perraud's *The Devil of Mascon*, in which a garrulous demon holes up in a French home, chiefly to taunt its inhabitants and anyone else who happens by over a period of weeks.

All of the above problems, with the exception of the last, will be with us for the remainder of this part of the book. The varied responses they receive from our philosophers will tell us much about the precise form of mechanism, if any, they hold.

[17] More (1662: i. 64 ff.).

16

Relations and Powers

If there are serious problems for ontological mechanism in England, its course-of nature cousin gets a powerful boost from developments in the theory of relations.

Recall that the Cartesian rejection of powers was built on a rejection of the view of relations that underpinned them. One clear way, then, to 'sanitize' powers—to make them acceptable within a mechanist ontology—would be to begin by treating relations and then move on to their subset, powers. And in fact this is precisely how Boyle approaches the issue in 'The Origin of Forms and Qualities.' It is surprising how often the importance of relations is missed; exploring the ontology of powers makes sense only against the background of a general metaphysical picture of relations.

What underpins this new demystifying approach to powers is its reductive view of relations. What precisely this means will emerge as we go. Put most simply: if relations in general, and powers in particular, can be shown to involve no ontological commitment over and above the qualities on the mechanist's preferred list, there is nothing for the Cartesian to balk at: no 'little souls' running about, and no intrinsic intentionality.

Course-of-nature mechanism can then be preserved, since events in the natural world are fixed by the powers of bodies, which in turn are nothing beyond the intrinsic qualities of those bodies. By taking a bottom-up position on laws of nature and at the same time avoiding the obscurity of Aristotelian powers or Cudworth-style 'plastick natures,' one might be able to reconcile course-of nature and ontological mechanism. The trick, of course, is to provide a sanitized concept of power. Once one does this, however, course-of-nature mechanism becomes inevitable: reduce powers to the intrinsic qualities of bodies, and nature takes the course it does simply because it is populated by bodies with those qualities.

Now, I will argue that Boyle is a *faux ami* where the project of course-of-nature mechanism is concerned; his positive position presents a paradox that can only be resolved by casting him much more in the mold of Descartes than, say, Hobbes, or even Locke. It is tempting to see Boyle as offering a sort of *via media* between the available positions; but there are simply too many *viae* on offer for this to be a useful thought. I shall argue that Boyle presents a *sui generis* position, with its own idiosyncrasies and difficulties. Thus, much of what I say in the following two sections will have to be qualified, since, as we shall see, the proper interpretation of Boyle is more complicated than one would have hoped. Nevertheless, his treatment of relations does, I think, provide Locke a jumping-off point, and is interesting in its own right.

And although the reductive view of relations does not in itself help to answer any of the conundrums set out in the previous section, it does provide a powerful incentive to look for some further strategy for dealing with them. For the alternatives prominent in the British context that are consistent with course-of-nature mechanism are unappealing, to say the least; indeed, they are (so far) exhausted by Cudworth's and More's absurd 'deputy gods.' In this chapter, I turn to Locke's and Boyle's arguments against the 'reality' (in a sense we shall have to discover) of relations. Whatever their positive views, they clearly share Régis's impulse to demystify powers by reducing them to something else, namely, the properties specified by ontological mechanism. To do this, however, they must purge relations of the *esse-ad* that the scholastics and their opponents hold is their defining feature. That is, Boyle and Locke must find a way to argue for the genuine polyadicity of relations while at the same time reducing relations to the intrinsic properties of the relata. I shall argue that such a view ultimately fails, since it cannot make sense of extrinsic relations. But this is a problem for their positive accounts, not for their arguments against the Aristotelian conception, and it is the latter that concern us now.

When Boyle introduces 'the primary affections' of matter in 'The Origin of Forms and Qualities,' he is careful to deny these 'moods' the status of real qualities. The point of calling them 'moods' (modes) is that size, shape, and either motion or rest are nothing over and above ways a given object has of existing. Real qualities, by contrast, are supposed to be 'real entities distinct from' the object, and capable of independent existence. The scholastics' mistake is to take the existence of a distinct name for a quality to entail the existence of a corresponding quality as an extra ingredient, as it were, in the subject. Thus, when we see a ball moving, its motion 'is not nothing, but it is not any *part*' of the ball.[1] The doctrine of real qualities accords these primary affections of matter the status of substances, and makes them accidents in name only. (This argument against real qualities is hardly novel.[2])

It is important to note that Boyle's arguments in this section of 'Origin' are directed primarily not at relational realism but at the doctrine of real qualities. Peter Anstey has forcefully argued that these passages are simply neutral with regard to the question of reduction.[3] What they challenge, on his view, is only the Suárecian claim that some features of objects can become detached from their subjects and serve as adjunct substances to that substance they would ordinarily

---

[1] OFQ 22.
[2] See Descartes's Letter to Elizabeth, May 21, 1643 (AT iii. 667/CSMK 219).
[3] See Anstey (2000, ch. 4).

be taken simply to modify. Whether relations or powers are reducible to the non-relational qualities of the beings that have them is, for Anstey, an entirely separate question.

Anstey is quite right to point out that the chief target of this portion of 'Origin' has often been missed.[4] A critique of powers as real qualities is not *ipso facto* an endorsement of reductionism. But while this connection is not so easily made, neither is it easily broken. For there are many ways of denying real qualities, and surely reductivism is among them. That is, one strategy for undermining real qualities is to argue that what the scholastics think are real qualities are not *qualities* at all but relations. This is meaningless, of course, unless one can distinguish relations as genuine two-place properties from qualities, and we have seen that scholastic relational realists do not do this: for them, a relational predicate stands for two distinct relational properties, each with its own directedness and status. Boyle proposes to carry out this project: relations come and go with the objects that have them and require no ontological support in the relata beyond their non-relational properties.

If I am right, reductionism about relations is Boyle's alternative to both relational realism and the doctrine of real qualities. It then functions in his argument against real qualities both as a replacement for that doctrine and as a diagnosis of it. For it is not as if the scholastics who went in for real qualities and relations were simply out of their minds; they recognized certain surface phenomena and thought real qualities were necessary to explain them.

As Boyle sees it, his opponents' original sin is their tendency to reify qualities.[5] What Boyle calls 'the grand mistake' is the supposition that for each quality we refer to, there is a distinct, corresponding quality, whether 'real' or not, in the object. And realism about relations—that is, the claim that relations constitute an irreducible, further feature of the world, beyond the objects so related—is one result of this mistake. Let us turn to one of Boyle's arguments against relational realism:

> It may be very useful to our present scope to observe that not only diversity of *names*, but even of *definitions*, doth not always infer a diversity of *physical entities* in the subject whereunto they are attributed. For it happens in many of the physical attributes of a body, as in those other cases wherein a man that is a father, a husband, a master, a prince, &c., may have a peculiar definition (such as the nature of the thing will bear) belong unto him in each of these capacities; and yet the man in himself considered is but the same man who, in respect of differing capacities or relations to other things, is called by different

---

[4] Anstey's targets include Peter Alexander (1985: 70–3), who focuses exclusively (in the context of this question) on 'Origins,' and who, on Anstey's reading, mistakes Boyle's attack on real qualities for an attack on relational realism, as well as Laura Keating (1993).

[5] Nor is this point unique to Boyle; as Galileo puts it in *The Assayer*: 'Just because we have given special names to these qualities, different from the names we have given to the primary and real properties, we are tempted into believing that the former really and truly exist as well as the latter' (in Matthews 1989: 57).

names, and described by various definitions, which yet (as I was saying) conclude not so many real and distinct entities in the person so variously denominated. (OFQ 22–3)[6]

With this argument, we move from an attack on real qualities to an attack on relational realism. (No one would have thought of fatherhood or husbandhood as a real quality.) And if we jump ahead a bit to Boyle's account of powers, we can see that this line of thought applies not just to real qualities but to the claim that distinct relations require distinct qualities, separable or not, in the object.

We may consider, then, that when Tubal Cain, or whoever else were the smith that invented *locks* and *keys*, had made his first lock . . . that was only a piece of iron contrived into such a shape; and when afterwards he made a key to that lock, that also in itself considered was nothing but a piece of iron of such a determinate figure. But in regard that these two pieces of iron might now be applied to one another after a certain manner . . . the lock and the key did each of them obtain a new capacity; and it became a main part of the notion and description of a *lock* that it was capable of being made to lock or unlock by that other piece of iron we call a *key*, and it was looked upon as a peculiar faculty and power in the key that it was fitted to open and shut the lock: *and yet by these attributes there was not added any real or physical entity* either to the lock or to the key, each of them remaining indeed nothing but the same piece of iron, just so shaped as it was before. (OFQ 23; last emphasis mine)

In this crucial passage, Boyle at one stroke disposes of the difficulties Descartes and others found in the ontology of power.[7] Like fatherhood, powers are relations. It is then a serious mistake to suppose that a given object could have a power intrinsically. On the scholastic view, powers, whether they are identified with real qualities or not, are had by objects even in total isolation from other objects. Powers are intrinsic properties. True, without the appropriate objects in the environment, a power will not be actualized; this happens only when the patient possesses the relevant passive power. Nevertheless, both the active and passive powers are intrinsic to the agents that have them. On this realist view, having a power to melt cheese is a quality just like being hot or being made of iron. Boyle clearly disagrees. Objects gain and lose capacities depending on the existence and arrangement of other bodies. This means those capacities are nothing over and above the primary affections of matter that characterize the relevant objects.[8] In just the same way, a nickel can be now 500 and now 1,000

---

6 Cf. CQ 287, where Boyle writes, '. . . I consider, that the Qualities of particular Bodies (for I speake not here of Magnitude, Shape, and Motion, which are the Primitive Moods and Catholick Affections of Matter it self) do for the most part consist in Relations, upon whose account one Body is fitted to act upon others, or disposed to be acted on by them; as Quicksilver has a Quality or Power (for here I take Qualities in the larger sense) to dissolve Gold and Silver, and a Capacity or Disposition to be dissolved by *Aqua fortis*, and (though less readily) by *Aqua regis*.'

7 The difficulties I have in mind here were covered above, in Ch. 5. This is not to say that other problems do not remain, as we shall see.

8 I do not mean to suggest that a thing's powers can be reduced to its texture. Given that powers are multilaterally reducible, an object can exist, whatever its texture, without having the power conferred on it by the existence of the other members of the supervenience base.

miles from Las Vegas without either the city or the coin undergoing any change in its intrinsic properties.

That this argument tells against not just the doctrine of real qualities but relational realism as well is easily shown. If Boyle were countenancing relational realism, one would expect him to at least mention that powers are features over and above the mechanical qualities of bodies, whether or not they are separable. In fact, he does just the opposite. Concluding his discussion of powers, he writes,

And this puts me in mind to add that the multiplicity of qualities that are sometimes to be met with in the same natural bodies needs not make men reject the opinion we have been proposing, by persuading them that so many differing attributes as may sometimes be found in one and the same natural body cannot proceed from the bare texture and other mechanical affections of its matter.   (OFQ 26)

It is not just that these qualities are not further *real* qualities of bodies; rather, they are not qualities at all, beyond texture and other mechanical affections.

Boyle's argument in both passages for the reducibility of relations is, I suggest, an argument from economy. The argument is explicit with regard to powers: 'Unless we admit the doctrine I have been proposing, we must admit that a body may have an almost infinite number of new real entities accruing to it without the intervention of any physical change in the body itself' (OFQ 24).[9] As it happens, this appeal to parsimony has a long medieval tradition, going back to Peter Aureoli. In the same vein, Aureoli writes, 'if the face of whatever man is different from the faces of all [other] men, the man's face will be burdened with innumerable realities, because he will have as many real dissimilitudes and differences of form in [his] face as in a subject as there are [other] men.'[10] And who wants his face burdened?

If we turn to Locke, we find a very similar line of thought.[11] Locke argues that relations are 'not contained in the real existence of Things' (*Essay*, II. xxv. 8: 322)[12] partly on the basis of economy. Locke writes, 'barely by the mind's changing the object, to which it compares anything, the same thing is capable of having contrary denominations, at the same time. *V.g.* Cajus, compared to

---

[9] In fact, this compressed argument also suggests the argument from change, examined below.

[10] Quoted in Henninger (1989: 156). For other sources, see, e.g., Ockham (1974: 162), where Ockham argues that if relations were real in this sense, there would be an infinity of entities in any given thing, given the relations of distance it would bear to the infinite parts that make up a material body.

[11] It has been suggested to me that the chapter on relations (II. xxv) is merely an episode in Locke's 'plain, Historical method,' and has no metaphysical implications at all. Locke's purpose, the objection runs, is simply to catalog our ideas, not to make claims about the world. The text itself makes this simply implausible. Not only does Locke make straightforward claims about the ontology of relations (e.g., II. xxv. 8: 322, quoted above), he also uses ontological claims in analyzing ideas themselves; for example, most complex ideas are not 'intended to be the Copies of any thing, nor referred to the existence of any thing' (IV. iv. 5: 564; examined below).

[12] For more on this passage, see Ch. 19. Note that I am interested at the moment in the arguments Locke and Boyle deploy; their positive views will be explored below.

several persons, may truly be said to be older and younger, stronger and weaker, etc.' (II. xxv. 5: 321).

This argument is clearly intended to support the anti-realist conclusion of II. xxv. 8. But how, exactly? On its face, it is difficult to reconstruct Locke's line of thought. Obviously, the 'denominations' Locke speaks of, when fully spelt out, will cease to be contrary at all: no one is older or younger *simpliciter*, but only older than *x*, or younger than *y*. The interesting point here, if there is one, is simply that relations shift according to the mind's activity. The ease with which the mind shifts from one relation to another, holding one relatum constant, is a sign that the mind is making an essential contribution to the 'reality' of these relations. What is more important, the relational realist has to hold that a single thing has an indefinitely large number of relational properties. But 'there is *no one thing*, whether simple *Idea*, Substance, Mode, or Relation, or Name of either of them, *which is not capable of an almost infinite number of* Considerations, in reference to other things' (II. xxv. 7: 321)

Locke has a further, and equally traditional, argument in his arsenal:

The *nature* therefore *of Relation*, consists in the referring, or comparing two things, one to another; from which comparison, one or both comes to be denominated. And if either of those things be removed, or cease to be, the Relation ceases, and the Denomination consequent to it, though the other receive in it self no alteration at all. *v.g. Cajus*, whom I consider to day as a Father, ceases to be so to morrow, only by the death of his Son, without any alteration made in himself.   (II. xxv. 5: 321)

This is not a mere off-hand remark, for Locke uses the same argument against Sergeant in his marginal notes to Sergeant's *Solid Philosophy*. Sergeant argues that Locke cannot account for the distinction between mutual and non-mutual relations, that is, the distinction between those relations that are such that each relatum is directed at the other and those relations that obtain only in one direction.[13] In the margin, Locke writes, 'What change does the father in the Indies suffer when his son is born in England?'[14] On Sergeant's view, it seems, a subject can undergo a real and intrinsic alteration simply in virtue of a change in a substance to which it is related. From Locke's point of view, this amounts to a deeply counter-intuitive kind of action at a distance. Note that this example tells only against a realism about mutual relations. If being a son were a non-mutual relation, the father would not undergo any real change; he would merely cease to be the proper subject of an extrinsic denomination, which is a Cambridge change. The point, of course, is generalizable against all mutual relations.

This argument against relational realism has its source in Aristotle. 'In respect of substance there is no motion, because substance has no contrary among things that are. Nor is there motion in respect of relation; for it may happen that when

---

[13] See above, sect. 3.2.
[14] Notes on Sergeant, included in the Garland Press facsimile of *Solid Philosophy* (Sergeant 1984: 254).

one correlative changes, the other, although this does not itself change, may be true or not true, so that in these cases the motion is accidental' (*Physics* 5. 2. 225ᵇ10–13). Whatever this means, such writers as Ockham, Harclay, and Aureoli took it as a powerful argument against relational realism.[15] It is hard to believe Locke does not fall into this tradition.

We can reconstruct the argument as a *reductio*, given some intuitively plausible premises:

1. If F is an intrinsic property of a, then when a becomes F, a undergoes a real change. (Definition)

2. Real changes can be brought about only by spatially contiguous agents. (Assumption)[16]

3. If F is a relation inhering in a, and there is a mutual relation G in b such that Fa. Gb, and at some future time t $\sim$Gb, then $\sim$Fa at t.

4. a and b are not spatially contiguous. (Assumption)

5. At t, a undergoes a real change. (from 1)

6. It is not the case that at t, a undergoes a real change. (from 2, 4)

Therefore, F cannot be an intrinsic property.[17]

On the view suggested by these texts of Locke and Boyle, powers are only misleadingly one-place properties; once we see their relational nature, their status as occult and irreducible disappears. Conversely, until we see this relational nature, they must remain odd creatures indeed, for it will look as if an object by its own nature points to something outside of itself.

In the face of this kind of argument, some Aristotelians might have bitten the bullet: the *esse-ad* nature of *potentiae* was, after all, a crucial feature of their conception of the natural world. By contrast, classifying powers as relations and then reducing relations to intrinsic qualities allows one to sanitize the notion of power and make it acceptable for those unwilling to countenance any properties in bodies beyond size, shape, movement, and their derivatives.

Before proceeding to assess this position on relations and powers (and to wonder whether either Boyle or Locke in fact holds it), we should pause to see just how it fares against the Cartesian barrage. How, exactly, does the reducibility of relations help to reconcile a bottom-up picture with ontological mechanism?

---

[15] See, e.g., Ockham (1974: 160): 'if a relation were a distinct thing, then when it first accrued to an object that object would have some new thing inhering in it; and consequently, it would actually be changed. But this is contrary to what the Philosopher says in the fifth book of the *Physics* where he claims that thing can gain a new relation without in the least changing.'

[16] We should note that even the scholastic realists accept this premise; indeed, Duns Scotus uses it to motivate his own realist position. See Adams (1987: i. 219).

[17] Here, I am taking 'intrinsic' as equivalent to 'monadic.' This argument is also contained in Boyle's claim that unless his own reductionism about relations and powers is accepted, 'we must admit that a body may have an almost infinite number of new real entities accruing to it without the intervention of any physical change in the body itself' (OFQ 24).

Recall Descartes's first worry from Chapter 5: what I called the problem of ontological independence. As Descartes reads the Aristotelians, the power of a key to open a lock is supposed to be an irreducible feature of it, one that cannot be exhaustively explained by its mechanical features. Boyle's reductive account of powers, however, means that his ontology need include nothing more than Descartes himself was willing to countenance. The point of reducing powers is precisely to bring a broadly Aristotelian method of thinking about the natural world in line with the ontology of mechanism.

But in calling powers 'reducible,' I do not mean to say that they can be eliminatively reduced. The reduction is on the model of water = H_2O, not witches = unpopular elderly women in medieval societies. On the first model, the term being reduced is retained: there is, after all, still water; on the second, the term being reduced is at the same time eliminated: there really are no witches after all. If we prefer, we can say that on the reductive view, a power supervenes on the relevant mechanical qualities of the objects it relates: fixing the mechanical qualities of a set of objects will fix their powers, and the existence of a power is nothing more than the existence of these objects and their qualities. It is vital to note that Boyle is not claiming that powers are bilaterally reducible, that is, reducible to the 'catholic affections' of the agent and the patient. For much turns on what other objects are in their environment. Revisiting his earlier example in 1686, Boyle writes, 'a Key may either acquire or lose its Power of opening a Door (which, perhaps, some School-Men would call its *aperitive Faculty,*) by a Change, not made in itself, but in the Locks it is apply'd to, or in the Motion of the Hand, that manages It' (NN 189).[18] Powers thus cannot be reduced bilaterally; how many other qualities will need to be taken account of will vary with the case.[19]

Reductivism also allows us to dispense with Descartes's second worry: the argument from explanatory impotence. While it tells against powers conceived as occult qualities, or as irreducible qualities, it in no way threatens Boylean powers. For these just are the sets of mechanical qualities that will be fundamental to the explanations of natural phenomena.[20] Intuitively, once we have seen the shapes of the key and the lock, we have also seen why the key is able to open the lock. There is nothing left to explain. Moreover, we have given our explanation in

---

[18] In CQ, Boyle makes a similar point: 'I have in the *Origine or Formes* touched upon this subject already, but otherwise than I am now about to doe. For whereas that which I doe *there* principally, (and yet but Transiently) take notice of, is *That one Body being surrounded with other Bodies, is manifestly wrought on by many of those among whome 'tis placed:* that which I chiefly in *This Discourse* consider is, the Impressions that a Body may receive, or the power it may acquire, from those vulgarly unknown, or at least unheeded Agents, by which it is thus affected, not only upon the account of its owne peculiar Texture or Disposition, but by vertue of the generall Fabrick of the World' (288). My thanks to Dan Kaufman for drawing my attention to this passage.

[19] See also OFQ 26–7.

[20] See esp. the preface to OFQ 13. It is important here that powers are being reduced to *sets* of mechanical qualities, and not to mechanical qualities *tout court.* A thing's mechanical qualities do not 'come and go' in the way its relational qualities do. Nevertheless, the set of these mechanical qualities can appear and disappear depending on the attendance record of their members.

terms of powers and *not* laws of nature without importing any irreducibly new element to our ontology or appealing to God's activity. Power attributions are promissory notes for mechanical explanations.

How does this view fare against Descartes's third worry, about physical intentionality? Attributing this kind of quasi-mental directedness is merely another result of the grand mistake. When we try to fold a power back into the intrinsic qualities of a single object, it becomes mysterious how it could be related to the states of affairs it will or would bring about. But if powers are relations, they are not to be counted among the real and intrinsic features of the object. There is a price to be paid for this, since, as the Tubal Cain passage makes clear, powers come and go depending on the relata, just like other relations. A world in which no thinking beings exist is one in which the relation *is smarter than* does not exist, either. But it is intuitively plausible to think that a key has the capacity to open a given lock even in the possible world in which that lock does not exist.

Boyle does sometimes say, in other contexts, that powers or dispositions would persist even in the absence of the relevant relatum. Edwin Curley suggests that in these passages Boyle is thinking of sortal powers, which would persist absent the relatum, as opposed to individual powers, which would not.[21] A sortal power is had by an object in virtue of its belonging to a given class, while an individual power is had by an object in virtue of its ability to act upon other individual objects. Even if it had a given world all to itself, this key would still belong to the class of keys that can open a given class of locks. Curley's proposal might well be correct. What counts for our purposes is simply that, so far as individual objects and not classes of objects are concerned, a power depends on the existence of the relevant relata.

We can now turn to (4), the worry that objects endowed with powers would somehow have to be inherently goal-directed. This feature of the scholastic view also goes missing in Boyle, for much the same reason as physical intentionality disappeared. Boyle is free to agree with Descartes that the only goal-directed behavior is initiated by minds, whether ours or God's.

On this view, then, powers are relations, and relations require nothing over and above their relata in order to obtain. To fully understand this, we need to contrast these relational qualities (for Boyle, like Locke, acquiesces in calling many mere relations 'qualities'[22]) with those genuine qualities on which they supervene. And here is where our story gets a bit complicated.

We need first of all to distinguish between the qualities and relations a body has on its own (call them Q1s and Q2s, respectively) and those it has in virtue of its relation to other bodies (Q3s).[23] To say that a macroscopic body has a certain quality intrinsically or on its own is not to say that that quality is not in

---

[21] Curley (1972: 446 ff.).   [22] See Boyle, IPQ 115.

[23] Note that I am not here making the distinction between primary and secondary qualities. Q1s and Q2s are both primary qualities.

the end relational. (This point will be crucial when we come to Locke.) As I read Boyle, what come to be called 'primary qualities' include not only non-relational, intrinsic qualities (Q1s), but also qualities that consist in the relation of the parts of the object itself (Q2s).

For example, consider one of Boyle's chief 'mechanical affections,' texture. Texture is what I am calling a Q2. While relational, it is a relation among the parts of a single given body. A particle does not have texture, for this arises only when a number of particles are arranged in a given way.[24] When several corpuscles 'convene together' to form an ordinary macroscopic body, their individual qualities (size, shape, bulk, and perhaps motion), plus the 'posture' and 'order' of these corpuscles, give rise to a further new quality, texture. Although Boyle calls posture and order 'new accidents,' he takes them to be reducible to what he calls 'situation,' or relative position. Situation, then, seems to be merely another relation, although it is distinct from Q3s by being internal to the object itself. For Boyle's reasoning about locks, keys, and poisons ought to apply at this level as well: just as nothing intrinsic to the key changes when it takes on its relation to the lock, so nothing about the individual particles changes when they are assembled in a given way.

Given this, Boyle's 'primary affections' of a macroscopic object will involve not just ordinary categorical qualities like shape but also relations among the object's parts. 'There is in the body . . . nothing real and physical but the size, shape, and motion or rest of its component particles, together with that texture of the whole which results from their being contrived as they are' (OFQ 31). The texture is real, although it is really nothing over and above the component particles.

Boyle's attack on relational realism embodies what I shall call the 'minimal claim': when a proposition of the form aRb is true, it does not entail that a and b possess (Aristotelian) relational properties F and G, respectively.[25] It is vital not to confuse this with the 'robust claim': when a proposition of the form aRb is true, it is true *only* in virtue of the non-relational, intrinsic properties of a and b.

So stated, the robust claim amounts to a statement of the bilateral reducibility of relations. This is, of course, far too strong, not only because not all relations are dyadic, but because some that appear to be dyadic are not. Whatever else Locke and Boyle believe, they clearly hold that many powers admit only of multi- and not bilateral reducibility. Boyle argues in 'Cosmical Qualities' that nearly any given event will depend not only on the qualities of the bodies obviously involved in it but also on those of bodies that might be far removed from the action. As Locke puts it, 'Things, however absolute and entire they seem in themselves, are but Retainers to other parts of Nature' (IV. vi. 11: 587).

---

[24] See OFQ 41.
[25] For more discussion of Aristotelian relational properties, see above, sect. 3.2.

Nor is this influence confined to the planets or other macro-level bodies; Locke thinks we are equally tempted to ignore 'the Operations of those invisible Fluids, [bodies] are encompassed with' (IV. vi. 11: 585). A better formulation of our robust claim, then, needs to allow for the possibility that the intrinsic properties of bodies other than a and b can be required in order for aRb to obtain. The point is that relational propositions are true in virtue of the intrinsic qualities of the relata, even when these include more objects than those that are directly mentioned in the proposition itself.

The minimal claim rules out the competing realist ontology of relations; the robust claim, by contrast, offers a positive account of its own. It says that the truth-maker for any relational claim is nothing more than the non-relational properties of the relata. Another way to put this is to say that the minimal claim is entirely consistent with the supervenience base of the relation including elements outside the objects so related, such as the activity of God. It says that there are no further qualities of the bodies related, other than their intrinsic, non-relational qualities. By contrast, the robust claim stipulates that relations supervene only on the intrinsic qualities of the relata.

Note that, while the robust claim entails the minimal, the converse entailment does not hold. Note also that, if powers are relations, the robust claim entails course-of-nature mechanism. I think it is beyond question that Boyle holds the minimal claim. His position on the robust claim, however, is anything but. Indeed, the key question of Boyle's overall view on the nature of the physical world and its powers is whether he makes this further commitment.

17

Boyle's Paradox

It is a core element of Boyle's approach to the natural world that powers are relations and hence require no new entities *in the bodies so related* beyond their intrinsic properties. This is just the minimal claim, and is supported by the passages we have looked at from his work of the 1660s. What is more, there are many grammatically monadic predicates that turn out to be polyadic in virtue of their ultimate nature as powers. Thus, Boyle writes, 'the qualities commonly called *sensible*, and many others too being according to our opinion but relative attributes, one of these now-mentioned alterations [e.g., a change in texture], though but mechanical, may endow the body it happens to with new relations both to the organs of sense and also to some other bodies, and consequently may endow it with additional qualities' (IPQ 115). A body that looked white can now appear blue, even under the same conditions; this change is nothing but a change in texture, which brings with it a new relation to the perceiver.

Coupled with Boyle's analysis of the lock and key, this looks very much like a commitment to both ontological and course-of-nature mechanism. Although powers are real, they are really something else: sets of monadic, intrinsic properties of bodies, the 'catholic affections' of matter. And once these are in place, what happens, happens in virtue of them.

What is more, Boyle shares Cudworth's rejection of the notion of a law of nature as a real feature of the universe capable of 'executing itself.' Boyle confesses he cannot understand 'how mere and consequently brute bodies can act according to laws, and for determinate ends, without any knowledge either of the one or of the other' (RD 159). '[A] law, being but a *notional rule of acting according to the declared will of a superior*, It is plain that nothing but an intellectual being can be properly capable of receiving and acting by a *law*' (NN 181). To talk of laws is to indulge in a misleading metaphor. This is not to say that Boyle does not himself do so; he admits that he will use 'many common phrases, which custom hath so authorized' as to make them almost unavoidable without tedious circumlocution (RD 160; see NN 181).

But, unlike Cudworth, Boyle thinks it is at least in principle possible to have a completed natural philosophy that would not resort to 'plastick natures' or Aristotelian powers. What makes this possible is precisely the purged and sanitized notion of power carved out in the early pages of 'The Origin of Forms and Qualities.' This represents a genuine advance toward the sort of view developed by Régis: one that would carry out the mechanistic reduction of powers and qualities while preserving course-of-nature mechanism. This is captured by Boyle's famous comparison of the natural world to the clock at Strasbourg and by his common references to the world as an 'automaton' (e.g., at OFQ 49).

Boyle thus appears well suited to the role of hero in this part of the book. The core notion of natural philosophy is power; not law; these powers are nothing over and above the properties picked out by ontological mechanism. Instead of banishing powers, like Descartes and Malebranche, he transforms them, and so rescues a core of Aristotelian insight that is otherwise lost.

Unfortunately, matters are not so simple. For Boyle often speaks of God's 'general and ordinary support and influence' (OFQ 41, discussed below). He goes much further than this, and suggests that, at the beginning of the world, God must 'settle the laws' of motion (OFQ 70). It seems that God must not only create and preserve matter and motion, he must also decide how motion is to be transmitted from one object to another. What is more, there seem to be no constraints on God's decree: 'the laws of motion . . . did not necessarily spring from the nature of matter, but depended upon the will of the divine author of things.'[1] This means that any general principles discoverable by natural philosophy owe their existence, not to the 'catholic affections' of matter, but to the will of God. Boyle writes,

If we consider God as the author of the universe, and the free establisher of the laws of motion, whose general concourse is necessary to the conservation and efficacy of every particular physical agent, we cannot but acknowledge that, by with-holding his concourse, or changing these laws of motion, he may invalidate most, if not all the axioms and theorems of natural philosophy . . .[2]

Such pronouncements create two sources of tension in Boyle's view. First, although he explicitly rejects talk of laws of nature (in NN and RD), and conducts many of his discussions in terms of powers rather than laws, in passages such as the above he helps himself to talk of laws, axioms, and theorems.

The real paradox here, however, is this: Boyle's emphasis on God's free establishment of the laws of motion is in direct conflict with his analysis of power. Consider again the lock and key analogy. If God can at will vary the amount of motion transferred in a given collision, the key's power to open the lock will depend, not just on the size, shape, and movement of the two bodies, but also on the amount of motion God chooses to allow the key to transfer to the tumblers. The same goes, of course, for all of the other examples Boyle gives: the power of crushed glass to poison, the power of the sun to warm, and so on. God's role as author and ubiquitous sustainer (by way of his 'general concourse') of the

---

[1] 'The Christian Virtuoso,' in Boyle (1772: v. 521).

[2] 'Reason and Religion,' in Boyle (1772: iv. 161). For further discussion, see Peter Anstey (2000, ch. 7).

laws of motion precludes any reduction of the powers of objects to properties of those objects themselves. That is, Boyle simply cannot endorse the robust claim above: he cannot consistently hold that the powers of objects are fully analyzable in terms of the non-relational qualities of those objects and their neighbors.

Consider the example of the lock and key once again. The Tubal Cain passage suggests that there is nothing more to the key's power to open the lock beyond the sizes and shapes of the two objects. But if motion is transmitted in some other way so that, say, it passes not from a moving body to the one with which it collides but to some distant body, the key will then no longer have the power to open the lock. Boyle allows that a world may be qualitatively identical to our own and yet

the laws of this propagation of motion among bodies may not be the same with those, that are established in our world; so that but one half, or some lesser part, (as a third) of the motion that is here communicated from a body of such a bulk and velocity, to another it finds at rest, or slowlier moved than itself, shall there pass from a movent to the body it impels, though all circumstances, except the laws of motion, be supposed the same.[3]

On this account, a key's power to open a lock cannot be identified with the intrinsic qualities of these objects, since they might possess all these qualities even when the laws of motion are such that the key cannot open the lock.

Let us take these problems in turn. Resolving the first tension is not as hard as it seems. What Boyle's official remarks indicate is that we must find some way of paraphrasing away talk of laws. We find a clue in 'About the Excellency and Grounds of the Mechanical Hypothesis,' where Boyle says that 'the laws of motion being settled and all upheld *by his incessant concourse and general providence*, the phenomena of the world are physically produced by the mechanical affections of matter' (EMH 139; my emphasis). It is, one way or another, through this concourse that the laws of motion are initially settled and continue to remain in effect. What this amounts to, I suggest, is simply that God creates and preserves the same amount of motion (and quantity of matter) in the world. In 'Origin,' Boyle writes that the naturalist 'has recourse to the first cause but for its general and ordinary support and influence, whereby it preserves matter and motion from annihilation or desition' (OFQ 70–1). To say, then, that God operates according to laws of motion is simply to say that he continues to create the same amount of motion. Talking of laws here is simply a disguised way of talking about how bodies behave; and their behavior is not fixed by their qualities *alone* but by the quantity of motion God chooses to conserve. Creating and conserving motion on one hand and 'settling and upholding' the laws of motion on the other are not two distinct activities but two ways of referring to

---

[3] 'Treatise on the High Veneration Man's Intellect Owes to God' (1684–5), in Boyle (1772: v. 140), also quoted in Anstey (2000: 175).

one and the same activity; and, given his pronouncements on the notion of law itself, we should regard the first as the more accurate.

It is important not to confuse laws of *motion* with laws of nature. In 'Cosmical Qualities,' Boyle includes the laws of nature among 'the catholic and unminded agents of nature,' along with 'divers invisible portions of matter' (CQ 289). Anstey takes Boyle to be thinking of laws, not as efficient causes, but as 'nomic' causes. This sits uneasily, however, with Boyle's hostility to the reification of laws. More importantly, it ignores the context of Boyle's use of 'laws of nature.'

Boyle has what I have been calling a 'summary' analysis of laws of nature, according to which they are nothing but ways of describing the *course* of nature.[4] What Boyle actually says in 'Cosmical Qualities' is that among the catholic agents of nature he includes 'the Establisht Lawes of the Universe, or that which is commonly called the *Ordinary Course of Nature*' (CQ 289). And this is borne out by his further use of the phrase in that work. For example, Boyle describes the behavior of a metal bar subjected to magnetic forces by lying in a given position through which, 'according to the establisht Lawes of Nature; the Magneticall Effluvia of the Earth must passe along' (CQ 295). This just means that, in the ordinary run of things, such effluvia tend to run in this direction. And in 'Cosmical Suspicions,' Boyle distinguishes between those uniformities that are most nearly analogous to laws and those that are more like customs, which admit of exception. This does not, of course, mean that Boyle holds a regularity theory of laws of nature. For these summary laws are themselves the result of more fundamental facts.

In fact, the laws (i.e., the patterns or course) of nature have a threefold dependence: on the laws of motion, on the powers and hence Q1s and Q2s of bodies, and on God's design and benevolence in creating and preserving both. Laws of motion thus sit at a more fundamental level than those of nature. The former are nevertheless regularities or uniformities; but, unlike laws of nature, they do not directly concern the behavior of ordinary objects. Instead, they concern the total amount of motion in the universe. Any event will be the product of two distinct efficient causes in cooperation: matter and God's activity as cause and sustainer of motion. Bodies do not act according to laws; bodies are acted on, by God (see RD 181–2), and this action of course depends on his willing to create, conserve, and propagate motion in regular and intelligible ways. When Boyle treats the laws of nature as causes, he means to appeal to them 'as an Helpe toward the giving an Account' (CS 305) of the qualities of things. As the example of magnetism shows, this help comes, not from finding another genuine cause, whether nomic or efficient, but from observing the settled behavior of other objects in the environment.

All of this sharpens rather than solves our paradox. God's role as creator and sustainer, not just of bodies and their initial arrangements, but of motion itself,

---

[4] For more on the two analyses of laws—conditional and summary—see above, sect. 1.1.

is the real cause of our second tension. The quantity of motion is conserved not because of its nature or that of bodies but because God chooses to conserve it. And if this is so, he can choose to conserve it in some different way, or not at all. Bodies thus cannot have powers, in the reductive sense we have been exploring.

But Boyle *does* think bodies have powers. How can we make sense of this? Perhaps Boyle simply changed his mind. But the reductive account of powers is suggested by 'Origin,' which includes the thesis that God must settle and enforce the laws of motion (OFQ 70–1). Alternatively, maybe Boyle is simply inconsistent, or never made up his mind; this appears to be the view of Anstey.[5] This is clearly a last resort.

What we need, then, is a way to reconcile the denial of powers as further elements in bodies with God as the creator and sustainer of motion. One possible solution would be to say that Boyle denies powers as 'further elements' only in the sense of real qualities. On this view, the passages from 'Origin' entitle us to say only that, whatever powers are, they do not have the same ontological status as substances. This does not mean that they are reducible to the 'catholic affections' of matter. And it is this attempted reduction that generates our problem.

True enough; but as we have seen, Boyle does take the extra step and deny that powers in bodies are irreducible; this is precisely how he argues against taking powers to be real qualities. To see our way through this, we need to distinguish two positions:

1. To say that *x* has a power to alter *y* is just to say that *x* has intrinsic property F, and that *y* has the requisite corresponding property G. (Compare the 'robust claim' above.)

2. To say that *x* has a power to alter *y* is to say nothing *about x and y* beyond their possession of intrinsic properties F and G. (Compare the 'minimal claim' above.)

Boyle clearly denies the robust claim, and so cannot hold (1) above. But that does not rule out (2); indeed, Boyle wants to affirm the minimal claim, so he must endorse (2). How can he do this?

If the key has the power to open the lock, it has it in virtue of the intrinsic features of the lock and the key *and* God's activity as the sustainer of motion. This does block the reduction of powers to the intrinsic properties of (pairs of) objects. But it does not entail anything at odds with the minimal claim. When we focus on either relatum of the power relation, we will never find anything but intrinsic, non-relational properties. These are not enough to endow bodies with powers. But the extra element that is required is not a further property of the body, beyond its mechanical affections. The extra element is the presence and continued transmission of motion. And this is what God adds.

---

[5] See Anstey (2000: 107–8).

Our second problem is thus solved.[6] The attribution of powers to bodies takes place only against the background of the divine sustenance of motion; this must be added as part of the reductive base of the power. None of this is at odds with Boyle's rejection of powers as irreducible features of the world.

It is tempting to think that, so far from fitting snugly into the hero's tights, Boyle is not so different from Descartes and Malebranche. All of them reject course-of-nature mechanism in order to preserve ontological mechanism. Boyle rejects Cudworth's 'plastick nature' in the name of ontological mechanism but in so doing acknowledges God as the source, not just of the initial quantity of motion, but of its continued existence *and* its habits of transference as well. It is no wonder that whenever Boyle talks about occasionalism, he has positive things to say about it.[7]

But this would be to underestimate Boyle's contribution to the debate. For, unlike Descartes and Malebranche, Boyle accords genuine powers to bodies, even if these depend in part on God's production of motion. By endorsing the minimal claim, as well as (2) above, and by pursuing his power-centered strategy, Boyle comes substantially closer to grounding scientific explanation in the properties of bodies and not the arbitrary will of a creator.

---

[6] This solution to Boyle's problem is in some ways similar to the interpretation of Locke offered by Edwin McCann. Although I argue against McCann's reading of Locke below, his ingenious proposal seems to work, with suitable adjustments, where Boyle is concerned.

[7] See *Royal Society Boyle Papers* (Bethesda, Md.: University Publications of America), vol. 10, fos 38–40 (late 1670s–1680s), in Anstey (2000: 210–13), discussed above, Ch. 15.

18

Boyle and the Concurrentists

Is Boyle a concurrentist? Like all problems of nomenclature, this one is intrinsically uninteresting: call him a concurrentist or not as you please; what counts is the content of his view. But its extrinsic value lies in helping us situate Boyle within the debate.

There is no doubt a family resemblance between Boyle and the concurrentists. Both have God and creatures collaborating to produce events. But the differences are more enlightening than the similarities.

To begin, note that the concurrentist sees God using creatures as instruments through which he accomplishes his goals, in such a way that one and the same event can be fully ascribed to God and to bodies. Boyle's view, however, is quite different; here, God and creatures seem to be genuine collaborators, with God doing one thing (creating, sustaining, and governing motion) and bodies another (deflecting or altering the course of that motion in accordance with their own properties). For this reason, Boyle has more in common with conservationism than concurrentism. His view is not all that far from Descartes's early one, in *Le Monde*. Contrast Régis, who would agree that God must create and sustain motion. But Régis's God only creates the substance of formal motion. On this view, governing the propagation of motion is left to bodies. On his view, then, there is still genuine sense to be made of bodies as instruments through which God works. For Boyle, by contrast, it is not enough that God create and sustain motion, along with the properties of bodies. He must go further and govern how that motion is distributed.

The difference between Boyle and the concurrentists is evident in their treatments of miracles. Consider the biblical example of Nebuchadnezzar's fiery furnace (Daniel 3). When Suárez discusses this example, he claims that God preserved Daniel's three companions by electing not to concur with the natural power of fire. Boyle does not account for the miracle in this way. Instead, God chose to 'suspend' the activity of fire by altering the laws of motion where the bodies of the intended victims were concerned.[1] God made it the case that the very fast motion of the particles of the fire did not transmit their motion to

---

[1] See 'Some Physico-Theological Considerations About the Possibility of the Resurrection,' in Boyle (1991: 207).

the bodies of the favored. This is at best an attenuated sense in which God can be said not to have 'concurred' with the power of fire.[2]

At the same time, neither is Boyle an occasionalist, despite his defense of occasionalism in an unpublished manuscript.[3] On one hand, his God does less than the occasionalists': Boyle's God need only create, preserve, and propagate motion, along with conserving bodies, while the occasionalists' God explicitly wills the position of every body in the world. On the other hand, Boyle's God must do more than Descartes's. Descartes claims that, even given a maximally chaotic initial position, the universe will resolve into order given the laws of nature, i.e., given God's adherence to a few simple rules. At the very least, Boyle's God must at a minimum 'contrive' the structures of created, and especially living, beings in an especially artful way at the very beginning (OFQ 70; RD 162). Such a claim is hardly controversial, in the modern period. But Boyle's God must also constantly decide whether to continue to propagate motion as he has always done. This is a free choice, and one he departs from whenever he chooses to perform miracles. Descartes's God, in virtue of his immutability, has no such choice to make.

Boyle thus represents a genuine alternative to all of the views on offer so far. What counts here is how far he falls short of a thoroughgoing mechanism, one which marries ontological with course-of-nature mechanism. That achievement, I shall argue, had to await Locke.

---

[2] Similar considerations apply to other tags commentators have entertained. We have already seen how inappropriate it would be to cast Boyle as a 'nomic concurrentist'; the same considerations apply to Anstey's preferred tag, 'nomic occasionalist.' Talk of laws of nature in this context can only be misleading. What God does in preserving the laws of motion is simply to conserve it and decide how it is to be distributed.   [3] See Ch. 15.

19

Locke on Relations

Boyle's basic strategy, as we have seen, is to sanitize powers by treating them as relations and then offering some version of a reductive account of those relations. That Locke takes the first step of this demystification is easily shown: after all, Locke says '*Powers* are Relations' (*Essay*, II. xxi. 19: 243).[1] The powers of gold are 'nothing else but so many relations' (II. xxiii. 37: 317); most of the simple ideas that make up our ideas of substances are powers, 'which [are] Relations to other Substances' (II. xxxi. 8: 381; see II. xxi. 73: 286–7). That Locke takes the second step, and indeed goes much further than Boyle by removing God's activities from the reductive base of power relations, must now be established.[2]

While Boyle endorses only the minimal claim, I shall argue that Locke goes further and argues for the robust claim. This will have ramifications for the rest of his view, including his position on powers, laws of nature, and his approach to mechanism generally. My basic line of thought in the following sections is this: Locke's ontology of relations constrains any reasonable interpretation of his views on these other topics. Lockean powers are relations; relations, I shall argue, are reducible to the non-relational qualities of their relata; thus, any story about Locke's commitment to course-of-nature mechanism, and in particular his talk of God 'superadding' powers to bodies, must square with this reductive picture. Given the brevity of Locke's chapter on relations, it might seem surprising that it should have such momentous consequences. But to decide the philosophical importance of a text based on its length is obviously silly. Moreover, if, as I shall argue, the position Locke defends would have been familiar to his readers (and taken for granted by many of them), it makes sense that he would spend little

[1] Here is the entire sentence, together with some preceding material: 'it is not one *power* that operates on another: But it is the Mind that operates, and exerts these Powers; it is the Man that does the Action; it is the Agent that has the power, or is able to do. For *Powers* are Relations, not Agents: And *that which has the power, or not the power to operate, is that alone, which is, or is not free*, and not the Power it self: For Freedom, or not Freedom, can belong to nothing, but what has, or has not a power to act.'

[2] By this, I do not mean to claim that God is no part, on Locke's view, of the explanation for *why* a thing has the powers it does. As we shall see, the evidence of design exhibited by the world is, on Locke's view, evidence of God's role in setting up matter in the complex ways he did. But this does not mean that, once the world is thus arranged, the powers of beings depend on God's *continued* intervention, as they do for Boyle.

time arguing for and stating it, as opposed to developing the views that follow from it, together with the other views he holds.

I shall begin by defending my attribution of the robust claim to Locke before moving on to explore the exact nature of his positive view. This should be easy to do: after all, Locke writes that relation is 'not contained in the real existence of Things, but something extraneous, and superinduced' (II. xxv. 8: 322). On its face, this says that relations are not real elements of the extra-mental world but instead are creatures of the mind. By contrast, Rae Langton has recently argued that, so far from endorsing the reducibility of relations, Locke believes they float entirely free of the relata and constitute a novel element in his ontology.

Langton distinguishes two theses Locke might have held:

> Irreducibility I: Not all relations or relational properties are unilaterally reducible.
>
> Irreducibility II: Not all relations or relational properties are bilaterally reducible.[3]

The claim made in (I) is simply that x's relations do not always supervene on the intrinsic properties of x. Leibniz held (or might have held) the contrary: sometimes (as in *First Truths*, for example) he speaks as if all of a thing's relations, including, for example, its being heavier than Bluto, are intrinsic properties of that thing. It is agreed on all hands that Locke does not hold (I); I know of no commentator who supposes that Lockean relations are intrinsic to a single thing. The claim made in (II) is much more bold: it says that at least some relations do not supervene on the intrinsic properties of the two objects related. There is something misleading, however, about the insistence on *bi*lateral reducibility: obviously, three-place relations are not bilaterally reducible. Langton, however, seems to take Irreducibility II to involve the claim that at least some relations do not supervene on *any* set of intrinsic qualities, no matter how numerous; thus I shall take (II) to involve the denial of multilateral reducibility. Otherwise, (II) would be trivially true where three-place relations were concerned. And as we have seen, both Locke and Boyle insist on the multilateral reducibility of powers.[4]

Why should we think Locke held (II)? Like me, Langton emphasizes the importance of II. xxv. 8: 322, where Locke claims that relations have no place in the 'real existence of Things.' But her gloss on this claim is puzzling. Langton announces that she 'will take it [the phrase "contained in the real existence of things"] to convey the idea of supervenience. Relation does not supervene on the intrinsic properties of things.'[5] This strikes me as forced. Langton is not arguing

---

³ (2000: 81).

⁴ Here it is important to keep in mind that, for Boyle, the reductive base for powers will include not just the 'catholic affections' of matter but God's activity as well. This is the main difference between Locke and Boyle on the question of reducibility.   ⁵ (2000: 80).

that relations are comparisons made by the mind but rather that relations are mind-independent entities that are not multilaterally reducible. That is, she reads the claim that relations are not real as the claim that they are so real they are an extra element in Locke's ontology.

Having supposed what I have argued is false, namely, that the denial of real existence to relations provides prima facie support for Irreducibility II, Langton constructs a devil's advocate to argue against it. As I shall show, the devil can do considerably better, and Langton's rejection of these arguments in no way tells against the reducibility of relations.

The advocate's first argument is that Locke's optimistic claims for the prospects of natural science (IV. iii. 25: 556) mean that 'truths about secondary and tertiary qualities will follow, Locke hopes, not from truths about the primary qualities of one thing considered alone [which is denied by Irreducibility I], but of two.'[6] But in the optimistic passage she refers to, Locke says nothing at all about secondary qualities: he says only that, given knowledge of real essences of bodies, 'we should know without Trial several of their Operations one upon another.' In the next passage she quotes (IV. iii. 25: 556), Locke speaks only of events involving primary and tertiary qualities, such as purging, killing, dissolving, and putting one to sleep.

Let us grant, however, the inclusion of secondary qualities in the scope of the reducibility thesis: on my reading, then, just as on the advocate's, the relation of causation between a piece of chalk and the quale it produces in us is nothing over and above the intrinsic qualities of the relata. What follows? Here Langton points us to the many passages in which Locke's pessimism is at the fore: he says that the ideas of secondary qualities 'can, *by us*, be no way deduced from bodily causes' (IV. iii. 28: 559; my emphasis). I shall examine others below, when I turn to the issue of Locke's mechanism. But what is notable about *all* of these passages is that they not only admit of but demand an epistemic reading. What counts is not whether *we* can deduce that the quale yellow will be tokened when we approach a school bus with such and such a microscopic structure, but whether there is in principle some micro-level primary quality that causes our quale. And what Locke says about relations entails that there will be. It does not entail that we know, or even can know, what it is. And for that reason, our attempted deductions fail. (In particular, it should be no surprise if the connection between bodily causes and qualia is mysterious, since, on Locke's view, we know almost nothing about the nature and ontological status of the mind.) If I am right, and the pessimistic passages are all epistemic in nature, then this objection to the advocate's position fails. As this line of objection is the chief focus of Chapter 21 below, I shall leave the matter there for now.

The advocate's second argument is that Locke's requirement that all relations have foundations in the relata entails the bilateral reducibility of relations. Against

---

6 Langton (2000: 85).

this, Langton has only to adduce the example of cassowary kinship, which is reduced to the fact 'that one laid the Egg out of which the other was hatched' (II. xxv. 8: 323). This is itself a relation that terminates, not in intrinsic qualities of relata, but in further relations. Here I agree with Langton. But I repudiate the argument. Moreover, contrary to appearances, Langton's refutation of this argument in no way tells against the thesis of multilateral reducibility. The fact that some relations have as their foundations qualities that turn out to be themselves relations is consistent with reducibility: so long as *at some level* the relations terminate in intrinsic qualities, the need for foundations, and for reducibility, will be met.

It is one thing to reject the doctrine of relational realism, where this involves a commitment to irreducibly relational properties. It is another altogether to come up with a replacement for it. Although the evidence for the claim that Locke holds the robust claim is persuasive, it leaves out a significant element of his view. From Locke's point of view, the robust claim is quite true, so far as it goes, but it neglects the contribution made by the mind. For as often as Locke suggests a reductive view of relations, he also suggests a conceptualist or anti-realist view, according to which relations and the facts they figure in are mind-dependent. Neither of these proposals is quite right; however, it will take some doing to see precisely how each goes wrong.

In fact, I think there are three distinct views suggested by Locke's texts, each with its own defects and virtues. Nevertheless, one view emerges as clearly superior, both philosophically and as an interpretation.

(a) *Reductionism.* Let us begin, not with the *Essay*, but with Locke's earlier shorthand notes on extension. There, he claims that the distance between any two particles 'is nothing but a mere relation which, though it be real, yet being the result from the being of other beings, requires no new act of creation, no more than it was necessary that paternity, which was not before in the world, should be created upon Adam's begetting Cain.'[7] The existence of paternity is nothing over and above a set of events that we choose to compare. In contemporary terms, we can take Locke's claim that a relation is 'the result from the being of other beings' as (reductive) supervenience: a and b's standing in relation R requires nothing more than the relevant qualities of a and b.[8]

---

[7] Bodleian, MS Locke, f. 1, p. 314, in Locke (1954). Note that distance, according to the *Essay*, is a simple mode of space rather than a relation; all modes, including such things as murder and gratitude, 'contain not in them the supposition of subsisting by themselves' (II. xii. 4: 165). Throughout the *Essay*, Locke often treats modes and relations together. My interest is squarely in relations as Locke conceives them; to what extent my comments apply to modes is a further issue. Moreover, in these notes, Locke also uses the example of paternity, which also recurs in the *Essay* (II. xxv. 2: 320).

[8] I say 'reductive' supervenience here because supervenience by itself is merely a thesis about the covariance of properties across possible worlds. This covariance obtains, on the present reading, *because* the relation is reduced to the relata.

I think there is a pretty straightforward problem with the reductive view. We can put it in the form of a dilemma: either the 'being of other beings,' the supervenience base, includes relations or it does not. If it does, the reduction is not complete. If it doesn't, the reduction cannot go through at all.

To take the second horn first: suppose we were to try to reduce relations to a non-relational set of properties or objects; call this view (a'). This might work with intrinsic relations, such as *lighter than*; this relation, one might think, is reducible to the tan car and the black car in the driveway. Once you have the two cars, you get *lighter than* for free, as it were. But clearly extrinsic relations cannot be reduced in this way. Suppose a is above b. This relation does not depend on a and b alone. So let us add that objects c, d, and e exist between a and b and account for their standing in this relation. But the further relations (among $c-e$) do not supervene on the objects or any of their intrinsic properties. Objects $a-e$ might exist in such a world with all their intrinsic properties intact and yet in a totally different spatial ordering. Nor will it help to say that among their intrinsic properties is a spatial location. Being-in-location-z is just another relation.

In fact, it seems to me that most reductive realists end up impaled on the other horn; call this view (a''). For example, Ockham holds that, as Marilyn McCord Adams puts it, 'statements of the form "*aRb*" assert no more than that a and b exist in a certain way, without implying the existence of a relation in any way distinct from the relata.'[9] This sounds quite level-headed, but it is not a *reductive* realism. For what is it to say that a and b 'exist in a certain way' other than that they stand in some relation? Analyzing the relation of fatherhood, Ockham himself holds that 'Socrates is the father of Plato' 'signifies only that Socrates has begotten Plato and both actually exist.'[10] But clearly begetting itself is a relation that requires some analysis. And we have already seen that the mere existence of two relata does not entail that they stand in that relation. So something else must be going on at the metaphysical level.

To sharpen this point, consider these questions:

1. Are relations real?
2. Is there any extra-mental fact that makes relational predications true or false?

Adams thinks Ockham can answer yes to (2) and no to (1). And of course Locke can as well, if (1) concerns relations understood in the Aristotelian's sense. But it is very hard to see how we could get a 'yes' to (2) without relations understood as further elements of the world that serve as the truth-makers for relational

---

[9] Adams (1987: i. 252). I would like to leave open the question whether Ockham is properly understood as a reductive realist in my terms or some kind of conceptualist; it seems to me that his texts oscillate between treating relations as names, as terms of the second intention and hence concepts, and as reducible to non-relational qualities.

[10] Quoted in Adams (1987: i. 252).

statements. That is, merely denying that relations are intrinsic properties doesn't let us see what makes it the case that a relation obtains. Some further positive story must be given.

(*b*) *Conceptualism.* This, on its face, is the official position of the *Essay*. Locke holds that the '*nature* therefore *of Relation*, consists in the referring, or comparing two things, one to another' (II. xxv. 5: 321; see II. xii. 7: 166). All relations 'terminate in simple *Ideas*' (II. xxv. 9: 233). In the relation of lighter than, for instance, the simple ideas, the foundation of the relation, will be the ideas of the colors of the relata. The termini or foundations of relations are just the simple ideas that fix the respect in which they are being compared. This is only the simplest kind of case, however, and Locke acknowledges that some relations terminate in a single complex idea (II. xxv. 1: 319). I suggest that Locke allows this kind of case to accommodate relations that terminate in other relations (which themselves, of course, are complex ideas), as in the case of cassowary kinship. In either case, the relation is a result of the mind's activity.[11]

On the conceptualist view, whether '*a*R*b*' is true or false depends on whether a mind is comparing *a* and *b* in respect R. Whether Cajus is a father or son depends on whom he is being compared to, and under what concept. Thus stated, the view seems to collapse into an extremely implausible form of idealism. (Indeed, I doubt whether anyone defended conceptualism so understood.) In worlds without minds, there just are no relations; Auckland is not south of London. And this is flatly inconsistent with Locke's realism about powers. On the other hand, it does accord well with his nominalism: since everything that exists, on his view, is fully particular, there cannot be any real entity in the world corresponding to '. . . is to the south of. . .' Such a relation would amount to a universal that admitted of multiple instantiations.

Nevertheless, I think conceptualism, so stated, cannot be Locke's view. If it were, his overall position would be deeply incoherent: he helps himself to talk of causal relations in his account of how simple ideas represent their objects (II. xxx–xxxii), the linchpin of his epistemology. Our ideas must really be caused by external objects for them to represent what they do, and this must hold regardless of whether there is any mind comparing objects and ideas under the concept of causation. And as we have noted, the qualities that make up our ideas of substances are, for the most part, not true qualities but 'nothing else but so many relations' (II. xxiii. 37), i.e., powers.

Thus, conceptualism faces a dilemma parallel to that posed above with regard to (*a*): is there a real feature of the world that justifies our application of relational concepts or not? If there is, then the ultimate status of relations is mind-independent; but then it is equally clear that the analysis of relations in

---

[11] Berkeley also states this view in his *Treatise Concerning the Principles of Human Knowledge*, §§89, 142.

terms of mental contents pulls no real weight. A further, realist ontology must be invoked. And Locke's claim that relations have no real existence seems to rule this out. If, on the other hand, there is no such real feature of the world, then Locke is committed to an implausible idealism that threatens to undermine huge swaths of his entire metaphysical picture.

Let us see what can be done to square the rejection of relations with the truth of propositions stating that they obtain. In other words, what we want is a way for Locke to answer no to (1), full stop, and yes to (2). To begin, we can note that, for Locke, no complex ideas, barring those of substances, are 'intended to be the Copies of any thing, nor referred to the existence of any thing, as their Originals' (IV. iv. 5: 564). They are '*Archetypes* of the Mind's own making' and as such '*cannot want any conformity necessary to real knowledge.*' It is not as if the truths involving these ideas are arbitrary or wholly irrelevant to experience, however. The mixed modes involved in mathematics, for example, are, like relations, 'archetypes without patterns,' and yet the mathematician can be sure that what he knows about those figures, 'when they have barely *an Ideal Existence* in his Mind, will hold true of them also, when they have a real existence in Matter; his consideration being barely of those Figures, which are the same, where-ever, or however they exist' (IV. iv. 6: 565). Thus, the fact that an idea is neither copied from experience nor intended to represent any real feature of things is no barrier to the objective truth of mental propositions in which that idea figures. In particular, ideas of relations can fit or conform to their (total) supervenience base, even though there is no extra element to justify their application.

I think the problem we have with Locke's story on relations is analogous to a problem we have, or should have, with his conception of truth. Locke holds that all verbal propositions inherit their meaning from mental propositions; these, in turn, require not merely ideas but an act of the mind, corresponding to either 'is' or 'is not' (III. vii). Truth presupposes these propositions, so, absent any mind to connect ideas with the copula, nothing can be true (IV. v. 2: 574). And of course there is nothing in the nature of things to correspond to these mental acts. So although 'there are trees' (or the mentalese for this) is, trivially, not true in a world without minds, it is hard to imagine that Locke would hold that a world denuded of minds is incapable of growing trees. But how can this be? Locke holds that a proposition's being true requires the ideas that make it up to be joined or separated; at the same time, there is every reason to think he would accept that *there are trees* is true even in a mindless world.

A reasonable answer here is that our intuitive urge to claim that '*There are trees* is true' in this particular world can be satisfied by the counterfactual claim that *if* there were minds in that world, they would, with suitable experiences in that world, justifiably form the proposition 'There are trees.' Indeed, something similar must be true even in the world as it is. For Locke holds that any proposition lacks a truth value when there is no one in fact thinking it, for the simple reason that propositions are generated by the mind. It remains the case, though, that a suitably oriented observer could form this proposition, and this is enough to keep Locke from an absurd form of idealism. Further evidence emerges, I think, if we look at Locke's position on immutable truths.

In his *Remarks upon Mr Norris's Books*, Locke argues against the Malebranchean claim that eternal and immutable truths hold in virtue of relations between ideas in the mind of God. Locke writes,

> 19. Truth . . . lies only in propositions. The foundation of this truth is the relation that is between our ideas. The knowledge of truth is that perception of the relation between our ideas to as it is expressed . . .

> 22. If no proposition should be made, there would be no truth nor falsehood; though the same relations still subsisting between the same ideas, is a foundation of the immutability of truth . . . in the same propositions, wherever made.[12]

Thus, without a mind to think it, '2 + 2 = 4' would not be true. It remains the case, though, that the ideas, whenever thought of, stand in a relation of equality that then compels the mind to assent to the proposition. In §22, Locke might seem to suggest that relations subsist between ideas even when those ideas are not present in any mind; all he actually says is that they 'still subsist' even when no proposition is formed that reflects them. Even this, of course, is problematic, given Locke's insistence on the mind-dependent nature of relations; like truth and falsity, they should depend on the activities of minds. I suspect that what Locke means here is simply that, the intrinsic nature of the ideas being preserved, the mind could find a justification or foundation for thinking of them as standing in a given relation. This of course is most clear in analytic cases; the ideas of a triangle and of a three-sided figure are, once brought before the mind, seen to be identical. The synthetic propositions of geometry also depend on the ideas that compose them, though in a very different way, as we shall see below.

What about contingent truths, such as 'Bobo lost his keys'? Given Locke's official definition of knowledge, they too must hold in virtue of the connections among ideas. My goal here, happily enough, is not to make out how precisely this part of Locke's account is to work. Most writers have found formidable difficulties in reconciling contingent claims, and especially those of real existence, with Locke's official definition of knowledge. The point of beating about these neighboring fields is to show that Locke does not shy away from the conclusion that truth and falsity are mind-dependent while not showing the slightest sign of endorsing any kind of radical idealism. I then argued that the best way to make sense of his position on truth and falsity is to invoke the notion of a (mind-independent) fitness of things (or of ideas), corresponding to the relations we rightly affirm of them. Now, Locke sometimes at least seems to hold that

---

[12] Locke (1823: x. 256–7); see also *Essay*, IV. i. 9: 529; IV. xv. 1: 654.

the relations among ideas that justify our propositions in the case of a priori knowledge can persist even in the absence of their relata, i.e., the relevant ideas. But this, I suggest, cannot be more than a loose way of speaking, since, whatever else one makes of II. xxv, that chapter shows that relations cannot hold in the absence of their relata. But this loose way of speaking does capture something important, namely, that in these a priori cases, the mind contemplating these ideas can, with justification, compare them in various ways and thus generate knowledge. We can then take this as our clue to Locke's view of relations generally.

If this is right, then Locke holds what we might call (c) *foundational conceptualism*: while relations are fully mind-dependent and have no real being, it remains the case that the mind-independent world provides a foundation (and a justification) for us to form the ideas of relations that we do.

And Locke does indeed speak of a foundation of relations. He argues that relations presuppose 'two *Ideas*, or Things, either in themselves really separate, or considered as distinct, and then a ground or occasion for their comparison' (II. xxv. 6: 321; see II. xxv. 7: 322). The relation *is* the comparison (II. xxv. 5: 321), and is in that sense mind-dependent; nevertheless, there are real grounds for the comparison in the objects or ideas considered. While the comparison is of course mind-dependent, the features in respect of which things are compared need not be.

Foundational conceptualism is also an apt term for Peter Aureoli's view:

without apprehension, there is not a greater conformity in act between two whitenesses than between a whiteness and a blackness. Indeed it is a contradiction in terms, since a conformity is not other than a distinct apprehension falling between the two whitenesses. So taking away the apprehension and having remain the conformity . . . is incoherent. Nevertheless, apart from all apprehension, two whitenesses are said [to be] more similar than a blackness and a whiteness on account of the near potency ultimately necessitating the cognitive power to have in act such an indistinct apprehension that is called 'similarity.'[13]

Locke, of course, does not use the terminology of potency, apprehension, and act. Aureoli's basic idea, however, should be clear: although '*aRb*' is not strictly true when there is no mind there to think of *a* and *b*, *a* and *b* together justify and, in Aureoli's view, even necessitate the formation of the proposition *aRb*.

We might note that this line of thought is most persuasive where intrinsic relations are concerned. It is no accident that Aureoli chose similarity as his example; once the two whitenesses exist, one gets their suitability for a judgment of similarity for free. It is much harder to see how the account will deal with

---

[13] Quoted in Henninger (1989: 168).

extrinsic relations. Suppose *a* is to the left of *b*; surely it is not merely the existence of *a* and *b* that makes it suitable for us to form the idea of their flanking the relation *to the left of.* One way or another, one wants to insist, they must in fact *be* in that relation in order for me to have a justification for thinking so. This is just the problem (*d′*) faced above.

Given that neither (*d′*) nor (*c*) can account for extrinsic relations, is there any reason to prefer one over the other? I think (*c*) provides the best overall account of Locke's view, since it makes sense of the mind-dependence of relations. Attributing it to Locke turns on applying the principle of charity. It gets us the result we wanted: a reading of II. xxv and Locke's other remarks on relations that respects his rejection of them as real beings and yet does not turn him into some kind of idealist. Moreover, it fits in nicely with his remarks about knowledge and propositions to form a coherent, if incomplete, whole.

Here we should pause to ask how well foundational conceptualism deals with the Cartesian worries set out above (Chapter 5). On this view, as on the robust claim itself, the problems of ontological independence and explanatory impotence are solved. Given that the extra-mental ground of a relation will always be the objects' intrinsic properties, they are not metaphysically mysterious; how precisely they will be invoked by an ideal, demonstrative science will be explored below. What foundational conceptualism adds to the robust claim is a more convincing response to the accusation of physical intentionality. For although the robust claim clearly avoids this problem, it fails to explain what was right about the objection. Powers do indeed seem, one way or another, to point to or direct the mind toward other objects beyond those that have them. The robust claim on its own cannot account for this; if powers are simply reduced away, the directedness of powers is revealed as an illusion. But a better view would explain, though not save, the phenomenon. How can this kind of 'pointing' be reconciled with a mechanist ontology?

This is precisely where foundational conceptualism is superior. As Locke says, in forming ideas of relations, the mind 'so considers one thing, that it does, as it were, bring it to, and set it by another, and carry its view from one to t'other' (II. xxv. 1: 319). The directedness of powers is still there, but it is explained entirely in terms of the mental acts that generate relations. This contribution of the mind, no different in kind from its activity in forming propositions or thoughts generally, is a vital ingredient in relations. It alone explains why thinking of one thing's power turns our thoughts to other objects.

In what follows, I shall call powers and relations 'multilaterally reducible,' even though this is slightly misleading, on the conceptual foundationalist interpretation. For it ignores the contribution made by the mind. Nevertheless, it remains the case that, where the extra-mental world is concerned, there is nothing to a relation beyond the intrinsic, non-relational properties of the bodies involved.

Recognizing powers as a subset of relations lets us answer a famous criticism lodged by Hume against Locke's genetic account of the idea of power. Hume writes,

Mr. Locke, in his chapter on power, says that, finding from experience that there are several new productions in matter and concluding that there must somewhere be a power capable of producing them, we arrive at last by this reasoning at the idea of power. But no reasoning can ever give us a new, original simple idea—as this philosopher himself confesses. This, therefore, can never be the origin of that idea. (*E* 7 n. 12 ; cf. *T* 1. 3. 14)

Locke takes power to be a simple idea; simple ideas cannot be generated by reasoning; thus, Locke is contradicting himself.

Had Hume considered Locke's chapter on relations and its sequel, however, he might have seen that this charge is unjust. All ideas of relations are complex (II. xii. 1: 163); thus, the idea of power *cannot* be the simple idea Hume accuses Locke of conjuring from reason alone. The idea that arises in us on repeated exposure to changes in the natural world is a result of the mind's activity of comparison, and not a simple idea at all. Reasoning does not produce a new simple idea of power. This is not to say that Hume has made an elementary mistake; Locke himself encourages confusion (see II. vii. 8: 131). But this confusion need not prevent *us* from recognizing this important aspect of Locke's position.

Given that ideas of powers are always complex, why does Locke call them 'simple'? The ideas that make up our complex ideas of substance are almost all of them ideas of powers (II. xxxi. 8: 381). Ideas of powers, then, can be said to be simple *relative to* our ideas of substances; that is, they are the elements from which these ideas are made and are in that sense 'simple' (see II. xxi. 3: 234).[14]

I do not mean to suggest, of course, that Locke's view, properly interpreted, would be acceptable to Hume. For Hume simply does not see how the qualities that found Locke's relations can give us any information about the course of events: 'Solidity, extension, motion; these qualities are all compleat in themselves, and never point out any other event which may result from them' (*E* 7. 8). On Hume's view, an ideally situated observer would still be at a loss when trying to predict or indeed explain events in the natural world.

A full consideration of Hume's objections will have to wait. For they come into focus only when we have a full grasp of Locke's account of causation and power. Ultimately, we shall find that the core of the disagreement lies in their widely different accounts of the necessity of causal connections. It is thus to Locke's account of causal necessity that I now turn.

---

[14] For more on Locke's position on ideas of powers, see Jacovides (2003).

20

Locke on Powers: The Geometrical Model

Locke's ontology of relations and powers points to an alternative to the occasionalists' cognitive model of causality. So far, though, this is all it has done; it remains to be seen just how Locke proposes to account for the tie between cause and effect. Foundational conceptualism offers the outlines of a story here: while there is no genuine, mind-independent relation of cause and effect, there is still an objective *ground* on the basis of which a suitably disposed mind will (and should) think of this relation. I shall call what explains this ground the 'geometrical model' of causation.[1]

Recall Malebranche's 'argument from elimination' in Chapter 10. There, Malebranche asked, 'What would [a] power be? Would it be a substance or a modality? If a substance, then it is not bodies that act but that substance which is in bodies. If the power is a modality, then there will be a modality in bodies which is neither motion nor shape.' Locke takes Malebranche's first point: a power cannot be a further feature of a body, distinct from its qualities, or else the body cannot itself be said to act. This comes out most clearly in his discussion of the will. Railing against the scholastic tendency to reify faculties, Locke insists that it is minds or persons, not the will, that have the power to act or not to act. '*Powers* are Relations, not Agents' (*Essay*, II. xxi. 19: 243). Similarly, it is the body that has the power, not a faculty or homunculus attached to it. Like other defenders of course-of-nature mechanism, Locke does not see the problem

---

[1] It is worth just noting Locke's explicit argument against another model of causation, the so-called 'influx' model, which I have not so far discussed, partly because it is, even among Aristotelians, a minority view. Still, it has its defenders in the period. John Sergeant, for instance, claims that 'What I conceive of *Causality* is, that 'tis the Power of Participating or Communicating some *Thing*, or some *mode* of Thing, to the Patient, which was before some way or other, in the Thing that *caus'd* it . . . that which is thus *communicated* is the *Real Ground* on which the *Real* Relation of the *Effect* to its *Cause* is founded' (Sergeant 1984: 255). In the margin, Locke writes, 'So fire that softens wax and hardens clay has some way or other softness and hardness in it?' (ibid.). The influx model entails that a given agent can have contrary properties at the same time and so cannot be correct. Now, the Aristotelian might well reply that effects depend not only on the active powers of the agent but on the passive powers of the patient (a distinction Locke himself draws in the *Essay*), so that one and the same agent can give rise to different effects. This move, however, seems unavailable on the influx model, since it depends on the numerical identity of the property supplied by the agent and then received by the patient; if the property somehow changes in the transaction, there's no sense to be made of the claim that it pre-existed in the agent.

with embracing the second horn of this dilemma. In fact, 'the different Bulk, Figure, Number, Texture, and Motion of [a body's] insensible Parts' (II. xxi. 73: 287) are the *only* cause we can imagine for the ideas bodies produce in us and for the changes they bring about in one another. Given Locke's foundational conceptualism, we can call these properties the supervenience base of the power relation, always keeping in mind that this base is not identical with the relation itself, which requires a contribution from the mind in the form of the activity of comparison.

As is well known, Locke claims that our clearest idea of *active* power is derived from the mind's reflection on its own operations. Locke has been read as making a concession to Hume when he admits that observing the antics of bodies gives rise in us only to the idea of passive power.[2] Having given up on deriving the idea of power from outer experience, so the story goes, Locke stubbornly hangs onto the inner experience of power. Hume then emerges as a consistent empiricist, and Locke as a mere halfway house on the way to a full-blooded denial of causal realism.

This is quite wrong. For denying that experience of bodies provides us with an idea of active power is hardly tantamount to claiming that it provides no idea of power at all. Power, Locke writes, 'is twofold, *viz.*, as able to make, or able to receive any change: The one may be called *Active*, and the other *Passive Power*' (II. xxi. 2: 234). In bodies, the power to initiate a change is the power to initiate motion. Bodies, as far as we can tell, simply do not do this; our experience of bodies does not provide us with 'any *Idea* of the beginning of Motion' (II. xxi. 4: 235).[3] So all bodies can do is transmit motion that they receive from some other source. This source will have to be a mind, whether finite or divine. But this presents no barrier, in principle, to seeing the physical world as a network of powers. Locke is far from agreeing with Glanvill's claim that 'causes are so connected that we cannot *know* any without knowing all.'[4] Not having 'a view of *Nature* while she lay in her *simple Originals*,'[5] and hence not knowing the ultimate origin of motion, cannot prevent us from knowing what will happen at the next moment. Moreover, a universe structured by passive powers is quite enough to provide Locke with a realist account of body–body causation, and more than enough to draw Hume's fire.

The natural world, then, is a complex of bodies that change by transmitting motion to one another. The ultimate origin of this motion is another matter; Locke clearly thinks God must have initially kicked things off (IV. x), though

---

[2] For a good discussion of this debate, see Coventry (2003).

[3] Given that my goal is chiefly to examine the debate over the physical world, we can happily ignore the complications of active powers altogether.

[4] Glanvill (1665: 154). Glanvill claims that 'we cannot know anything in Nature without knowing the first Springs of Natural Motions,' by which he means the '*true initial causes.*'

[5] Ibid.

there is no suggestion that he thinks God must continue to act to conserve or perpetually reproduce motion.[6]

To see how the intrinsic qualities of bodies constitute their powers, we must of course understand just what those qualities are. I shall argue that Locke's taxonomy of qualities mirrors that of Boyle, which lets us clear up a common difficulty in understanding Locke's view.

Like Boyle, Locke accepts that some of the qualities of bodies, though they would have them even in isolation from other bodies, are really relations (Q2s). And these relations turn out to be fundamental to their powers: 'the Active and Passive Powers of Bodies, and their ways of operating, consist . . . in a Texture and Motion of Parts' (IV. iii. 16: 547). For any given macroscopic body, its powers will be reducible to the situation or texture of the particles of which they are formed and the motion of these particles relative to each other plus, of course, the texture and motion of the particles of the other bodies to which they are related. Locke also implicitly distinguishes between Q2s and Q3s, that is, between relations between an object's components and relations between that object and others. 'The *simple* Ideas whereof we make our complex ones of Substances, are all of them (bating only the Figure and Bulk of some sorts) Powers' (II. xxxi. 8: 381). Locke is not merely reiterating his familiar point that many of the constituent ideas of substances are ideas of secondary qualities and thus of powers; even primary qualities, barring figure and bulk, are powers and thus relations.

If this is right, we can bring the distinction between primary and secondary qualities into sharper focus: just as for Boyle, primary qualities include Q1s and Q2s, while secondary and tertiary qualities are Q3s. Locke calls secondary qualities *nothing but* powers, implying that primary qualities are powers *and* something else (II. viii. 24: 141; but see II. viii. 8). Having noticed the lock and key passage in Boyle, Reginald Jackson proposed that the primary–secondary distinction is meant to hold between categorical properties and the relational and dispositional states that supervene on them.[7] E. M. Curley has argued that this will not work, however, for Jackson's reading 'requires situation not to be a primary quality.'[8] Still, Jackson nearly had it right. If we apply the distinction we extracted from Boyle between Q1s and 2s (i.e., non-relational and relational qualities internal to a given object) on one hand and Q3s on the other, we can say that situation or texture is still a primary quality of a body, because its relata are parts of that body itself. On my view, a primary quality is a quality, whether relational (Q2) or not (Q1), that an object has on its own, regardless of the

---

[6] The issue is complicated by IV. iii. 29: 560, where Locke says that such things as 'the Resurrection of the dead, the future state of this Globe of Earth . . . are by every one acknowledged to depend wholly on the Determination of a free agent,' namely, God. I read this, not as the claim that God must continue to produce motion at every moment in the physical world, but as the claim that the future of the world is, in the broadest and least controversial sense, in God's hands.

[7] Jackson (1968).   [8] Curley (1972: 443).

arrangement and population of other items in its world. A secondary quality like a color, however, relates the object to others, in particular, to perceivers. Secondary, like tertiary, qualities must take as their relata at least two distinct macro-level objects; primary qualities are either non-relational or relate only the constituents of a single object.[9]

What is more, our analysis of Q3s as powers and hence as multilaterally reducible clears up a persistent source of puzzlement for Locke's readers. For Locke seems to think that powers come and go with a thing's circumstances, whereas, if we take powers to be dispositions, a thing ought to have those dispositions regardless of the circumstances that would, as it were, activate them. By contrast, the multilateral reducibility of powers makes them dependent on the circumstances of the bodies that have them in a very strong way. This is why Locke can say that if one were to 'Put a piece of *Gold* any where by it self, separate from the reach and influence of all other bodies, it will immediately lose all its Colour and Weight, and perhaps Malleableness too' (IV. vi. 11: 585). If gold is disposed to cause in us the idea of yellow, won't it retain this disposition even if it is put somewhere 'by itself'? But this is not Locke's position. Since secondary qualities are merely powers, they fail to obtain when the relevant qualities of their relata are not present.[10] Nor is this peculiar to secondary qualities: even weight and malleableness would disappear in such an environment. As with Boyle, the logical structure of ascriptions of power is polyadic, not monadic.[11] And polyadic properties obtain only when their constituents are present.

With this as background, we are now in a position to see precisely how Locke thinks cause and effect are connected. Let us begin by looking at the kind of completed science of the natural world Locke thinks is in principle possible. Like the Aristotelians, he holds that such a science would be demonstrative. That is, it would start with a priori necessary truths and, by entering the facts of the particular case (e.g., which bodies are situated where, and with what microstructures), logically imply the state of affairs that would result. For example, Locke writes that,

---

[9] Throughout, I mean by 'object' a macroscopic object, not a corpuscle.

[10] Contrast this position with George Molnar's. Molnar holds that the physical intentionality that ties a power to its effects cannot be a 'real' or 'genuine' relation. As Molnar puts it, 'the nexus between the intentional state and the object to which it refers *is not that of a genuine relation*. In the case of a genuine relation, for example a causal relation, all relata must exist. Not so with intentionality' (2003: 62). It is easy to imagine Locke and Boyle insisting that, if the tie between power and effect is anything, it is a genuine relation in Molnar's sense, i.e., all relata must exist for that relation to obtain. And on their behalf, it is hard to know what to make of a non-genuine or quasi-relation, which Molnar's view requires.

[11] *Pace* Alexander (1985: 165 ff.). Alexander distinguishes between powers in the epistemic and ontological senses. The former, he grants, are relational; but he thinks that Locke takes powers as they exist in bodies to be 'intrinsic and non-relational' (166). This is problematic, however. We have already seen many passages (such as IV. vi. 11: 585) in which Locke is speaking of powers not merely as we conceive them, but as they are in themselves. In the isolation test of IV. iv. 11, for example, it is clear that powers considered in things are nevertheless relational.

If we could discover the Figure, Size, Texture, and Motion of the minute Constituent parts of any two Bodies, we should know without Trial several of their Operations one upon another, as we do now the Properties of a Square, or Triangle . . . The dissolving of Silver in *aqua fortis*, and Gold in *aqua Regia*, would be, then, perhaps, no more difficult to know, than it is to a Smith to understand, why the turning of one Key will open a Lock, and not the turning of another.   (IV. iii. 25: 556)

In principle, then, someone who knew the real essences of bodies x and y (and whatever other bodies were concerned, including those 'invisible Fluids') could infallibly infer how x and y would behave. To do this, one needs to know not *just* those real essences, but the truths that govern their actions.[12]

What are these necessary truths? Unsurprisingly, they are the truths of geometry. These, plus the shapes of the lock and the key, are all that are required for knowing how one might be used to act upon the other; similarly, the size, figure, bulk, etc., of the insensible parts of bodies are the ultimate cause and explanation of their antics.

Here is the payoff for having struggled through Locke's foundational conceptualism. There, we saw that, like the ideas of mathematics, ideas of relations can be justifiably applied, even though they are not copied directly from experience. So even though we construct the ideas of power and causation, that the world sometimes answers to them can in principle be ascertained. Indeed, at a very basic level, simply witnessing changes in the world tells us that there are at least passive powers. Under ideal epistemic conditions, we would know what explains these macro-level events. What is more, in neither case would our knowledge depend on what objects the world actually contains. To know the properties of gold, 'it would be no more necessary, that *Gold* should exist, and that we should make Experiments upon it, than it is necessary for the knowing the Properties of a Triangle, that a Triangle should exist in any Matter' (IV. vi. 11: 585). In this sense, a completed Lockean science would be a priori. Locke's comparison between the properties of gold and those of a triangle is no accident: both reflect his commitment to conceptual foundationalism.

Moreover, in speaking of the properties of gold and a triangle in the same breath, Locke is drawing our attention to a further connection. He is using 'properties' in the traditional sense, that is, to refer to features of an object that, while not constitutive of its essence, necessarily flow from that essence.[13] So although it is not part of the definition of a right-angled triangle that it

---

[12] Note that Locke says only that we should know *several* of the bodies' operations on one another. Why does he hesitate here? The answer, I think, is to be found in the way Locke sets up the counterfactual situation: if we knew the intrinsic qualities of any *two* bodies, we'd be able to predict some, but not all, of their operations. Some, but not all, because, as I point out above, knowing *all* would require knowing the intrinsic qualities of not just two bodies but all the other relevant bodies.

[13] These *propria* are not to be confused with the powers that, on the scholastic view, are themselves contained in the essences of the thing. For more on this, see Ayers (1991: ii. 21 ff.).

obey the Pythagorean theorem, one can nevertheless deduce this from the idea of that triangle. Similarly, the physical object answering to the name 'gold' has a real essence, a certain arrangement of microphysical parts, that would in principle allow us to deduce its behavior (assuming we knew enough about those other objects in the world with which it interacts). In other words, the logically necessary connections that would be captured by an ideal science are, in contemporary terms, synthetic.

To see this, consider a well-known passage from the *Essay*:

> We can know then the Truth of two sorts of Propositions, with perfect *certainty*; the one is, of those Trifling propositions . . . And secondly, we can know the Truth, and so may be *certain* in Propositions, which affirm something of another, which is a necessary consequence of its precise complex *Idea*, but not contained in it. As that *the external Angle of all Triangles, is bigger than either of the opposite internal Angles*; which relation of the outward Angle, to either of the opposite internal Angles, making no part of the complex *Idea*, signified by the name Triangle, this is a real Truth, and conveys with it instructive *real Knowledge*. (IV. viii. 8: 614)

This goes to the heart of the geometrical model. Unlike the cognitive model, it does not presuppose that the tie between cause and effect is analytic. It is not as if a Lockean real essence just includes in its very nature the information that it will behave in such and such a way. There is no way it could, since powers are relations, which are multilaterally and not unilaterally reducible. This last point is one of the most important, and least often grasped, differences between Locke's real essences and those of the scholastics.

Recall that a scholastic power 'points to' its possible effects. Given their position on relations, the scholastics are forced to collapse powers into monadic properties. This is one common source, from the modern period to our own, of puzzlement: it is very hard to see how a single feature of an object could 'point' beyond itself, not just to other features of the actual world, but to non-actualized possibilia. By telescoping powers into genuinely two-place relations, Locke avoids this Meinongian predicament. Echoing Boyle's lock and key discussion, Locke suggests (IV. iii. 25: 556) that the geometrical model can ground genuine necessities in bodies without invoking the intentionality of powers or of God. But just how *are* they to be grounded?

The lack of adequate ideas of real essences prevents us from having the kind of demonstrative science Locke envisions. This is because our ideas of bodies, while they justify us in forming many ideas of relations among them, do not show us all of the *necessary* relations and connections (IV. iii. 16: 548). This is the recurrent theme of book IV: having no ideas of the 'particular primary Qualities of the minute parts of [sage or hemlock], nor of other Bodies which we would apply them to, we cannot tell what effects they will produce' (IV. iii. 26: 557). To be sure, we do detect some necessary connections among the primary qualities of bodies, e.g., that figure presupposes extension. But these are either trifling propositions or perilously close to them. Informative necessary connections between substances and powers must elude us as long as we lack ideas of real essences, for the simple reason that these connections are relations among the *insensible* parts of bodies.

Thus, while Lockean real essences perform almost none of the functions of Aristotelian forms—they do not serve as migration barriers;[14] they do not as a matter of analytic truth contain the powers of the bodies that have them—they are similar to Régis's replacement-forms in that they are second-order properties of bodies. These properties of the modes of bodies can in principle serve as premises in an Aristotelian demonstration. Locke is at once the first thoroughly modern empiricist and the last Aristotelian.

---

[14] Possession of a fully realized Aristotelian essence prevents a being from acquiring a new such essence without being destroyed. Lockean real essences do not impose any such constraint; indeed, it is an article of faith in mechanism that in principle any body can switch kinds by having its microstructure altered.

21

Locke's Mechanisms

We can now formulate ontological mechanism to fit with Locke's own hierarchy of qualities: the properties of bodies are exhausted by their non-relational qualities (Q1s), plus whatever internal relational qualities (Q2s) one chooses to include. The reducibility of powers to Q1s and Q2s, via conceptual foundationalism, is a vital part of the ontological mechanist's project, since only by this means can talk of powers be made innocuous. Does Locke also think that bodies behave as they do in virtue of these Q1s and 2s? That is, is Locke also a course-of-nature mechanist?[1]

The two questions are intertwined. It should be no surprise that whether Locke is in the end an ontological mechanist turns on whether he is also a course-of-nature mechanist. After all, Boyle's reductive account of powers was ultimately thwarted by his insistence on the role of God in fixing the course of nature. It is important to be clear on this point. Course-of-nature mechanism is perfectly consistent with the deistic view that God must come on the scene to give the material world its initial quantity of motion. In fact, this is precisely the view Locke's argument for God's existence suggests (IV. x).

Moreover, course-of-nature mechanism is consistent with the claim that the powers of bodies do not derive from the minute parts that make them up. It is, in other words, consistent with the falsity of corpuscularianism. It is *not* consistent, however, with the claim that it is possible for God to alter the course of nature by simply changing the laws of nature or motion. For this would mean altering the powers of objects without altering their reductive base. Powers would then have to have an existence beyond that of the base, and, since they are relations among bodies, this is impossible.[2]

Leibniz makes precisely this point when he argues in the *Nouveaux Essais* that 'if God were to give things accidental powers detached from their natures, and thus inaccessible to reason in general, this would be a backdoor through which to re-introduce those too occult qualities, which no mind can understand.' Contra Leibniz, Locke will have no truck with those 'little goblins' who 'come

---

[1] The interpretation I ultimately argue for is similar to Michael Ayers's. My chief argument for this interpretation is not, particularly since it takes as its starting point the ontology of relations.

[2] I consider below a voluntarist reading, that of Edwin McCann, which does not have this consequence.

forward like Gods on the stage' and do whatever the philosopher likes.[3] This claim is controversial; many commentators now suppose that Locke in fact is up to his neck in non-mechanical qualities or the constant intervention of God.[4]

My argument in this chapter is simple: Locke's ontology of powers and relations constrains any interpretation of his views on course-of-nature mechanism. In particular, the reducibility of relations prohibits the addition (or 'superaddition'[5]) of powers that are not founded on the internal properties (Q1s or Q2s) of objects.

An appealing thought here is that, since God can alter the course of nature, powers must not be reducible in the sense I have specified. Similarly, one might argue that since some qualities such as gravity and color cannot be explained by mechanical means, they cannot be reduced to any base in bodies at all. This is the kind of strategy exploited by Curley in arguing against Reginald Jackson's view, which, as I have noted, is similar to my own.[6] The basic argument is this: since Locke is a Cartesian voluntarist, powers are not reducible to non-relational qualities. But one man's modus ponens is another man's modus tollens.

We have already examined Locke's ideal of a completed science (IV. iii. 25). Were our epistemic state improved, our ordinary conceptual equipment of powers and their effects would suffice and remain unchanged, only the perspicacity of our senses having been altered. The most astute investigation would never reveal a third thing, an irreducible power or force; instead, our spade would have turned precisely where it should, for someone committed to both course-of-nature and ontological mechanism: on the intrinsic qualities of the objects we compare under the concept of power.

[3] *Nouveaux Essais*, in Leibniz (1978: 363).
[4] See, e.g., Wilson (1979), Stuart (1998), and Langton (2000).
[5] In what follows, my analysis of Locke is restricted to the non-human world. The terminology of superaddition is used in the *Essay* only in the context of the relation between thought and matter. Thus, I have chosen not to frame the debate in terms of the possibility of superaddition. It is worth noting, however, that Michael Ayers (1981: 228 ff.) provides a plausible reading of the superaddition passages that is consistent with my claim that Locke is both an ontological and a course-of-nature mechanist. Briefly, Ayers points out that, on Locke's view, God must also 'superadd' motion to matter (see the Second Reply to Stillingfleet (Locke 1823: iv. 461)). Motion is hardly a supernatural quality, or one that conflicts with the essence of matter. Thus, the claim that thought *might* be superadded to matter by God (IV. iii. 6: 541) need entail nothing more than that God acts on matter in such a way as to endow it with an accident that is, nevertheless, in no way contrary to its nature, any more than motion is contrary to the nature of matter. Downing (2007) offers a different view. She thinks that the possibility of superadding thought to matter is the possibility of God's adding a non-material element to an otherwise material being. Since superaddition concerns our nominal essences (i.e., the chance that there is something in body that is not contained in our *idea* of body), this should be no surprise, she thinks, since Locke holds that gravity already reveals the presence in body of something that is not contained in our idea of it. Downing's view is obviously friendly to my own, since it in no way implies that God could add powers to bodies without changing what it is that makes them up. I address the superaddition of non-mental qualities to bodies below.   [6] See Curley (1972: 450) and Langton (2000).

In the same passage, Locke claims that, if we were in this epistemically privileged position, 'The dissolving of Silver in *aqua fortis*, and Gold in *aqua Regia*, would be, then, perhaps, no more difficult to know, than it is to a Smith to understand, why the turning of one Key will open a Lock, and not the turning of another' (IV. iii. 25: 556).[7] The crucial element here is the suggestion that, after we have inspected the primary qualities of the corpuscles, there is no residue, as it were, left unexplained. God and his activities are simply not in the picture. Our understanding of the course of nature is complete in itself.

This way of putting matters, however, ignores Locke's mitigated skepticism about the explanatory powers of corpuscularianism. And of course it is to these explanatory limitations that the proponents of the voluntarist reading of Locke appeal. In his 'pessimistic' moments, Locke finds the corpuscular model wanting. Locke's pessimism is limited to four issues: cohesion (IV. iii. 29: 559–60), gravity (1823: iv. 467–8), impulse (II. xxiii. 28: 311), and the means by which primary qualities produce ideas of secondary qualities in us (IV. iii. 13: 545). I have already shown that by Locke's time, these were recurrent trouble spots for ontological mechanism. Now, on the voluntarist reading, Locke is claiming that ontological mechanism fails, and thus the reductive interpretation of powers (and hence course-of-nature mechanism) fails with it. For as Lisa Downing has noted, 'the [lock and key] analogy breaks down once cohesion and impulse are thrown into question.'[8] If cohesion and impulse remain mysterious, it seems there is, after all, an explanatory gap between the primary qualities invoked by corpuscularianism and the movement of the key in the lock.

We can see this gap most clearly in the problem of impulse, which I have called the problem of transference. Locke sometimes speaks as if a real property were transferred when objects collide. When one ball strikes another, 'it only communicates the motion it had received from another, and loses in it self so much, as the other received' (II. xxi. 4: 235). At first glance, it looks as if Locke might be claiming that a quantity of motion, considered as a numerically identical 'real quality,' is changing hands, passing from one ball to another. This seems odd, since simple ideas and the qualities they represent cannot exist apart from the substances they modify (II. xxxi. 1: 295). In fact, although Locke does say that transference is part of our ordinary conception of motion, he thinks such a supposition is unintelligible. Our only conception of impulse is 'the passing of Motion out of one Body into another, which, I think, is as obscure and unconceivable' as our conception of mind–body interaction (II. xxiii. 28: 311). As I read the passage, Locke is saying that our ignorance of the microphysical structures of the objects involved in motion means that we must fall back on a vague, metaphorical notion of transference, *not* that there really is, in a literal sense, a transfer of a single real quality. The notion

[7] Alexander (1985: 73) notes that Locke's examples in IV. iii. 25 mirror those of Boyle in OFQ.
[8] Downing (1998: 409).

of transference is a placeholder for a more intelligible notion *scientia* would provide.[9]

Downing suggests that corpuscular mechanism functions primarily as an illustration of the *kind* of explanation Locke thinks appropriate in science. Far from insisting dogmatically on corpuscularianism, Locke instead takes mechanism to be the best available hypothesis and the closest approximation of 'what it would be like'[10] to have scientific knowledge that traces out necessary connections and funds a priori inferences to future effects in the way indicated by the passages above. The crucial point is that, whether the corpuscular hypothesis is in principle incapable of closing these explanatory gaps or whether these gaps are *merely* the result of our impoverished access to microphysical qualities, Locke clearly thinks some such hypothesis is correct. And given his views on relations, we can say that, *whatever* that hypothesis is, and whatever particular primary qualities (Q1s and Q2s) turn out to be ultimate constituents of the universe,[11] a completed science will not appeal to anything besides those qualities; in particular, it will not appeal to irreducible relations.

To be clear: if the voluntarist claim amounts merely to the claim that the qualities on mechanism's preferred list of fundamental explanatory principles might not be sufficient to explain all the phenomena, then I am fully in agreement with it. As Locke says,

The gravitation of matter towards matter, by ways inconceivable to me, is . . . a demonstration that God can, if he pleases, put into bodies powers and ways of operation, *above what can be derived from our idea of body, or can be explained by what we know of matter*, but also an unquestionable and everywhere visible instance, that he has done so.[12]

To be sure, almost no powers are deducible from the contents of our (current, imperfect) idea of body. The reference to God here need import nothing supernatural but rather our ignorance of the natural.[13] But if the claim is that, over and above *any* intrinsic qualities things have, God can add to them new relations (including powers), it conflicts with Locke's position, since these have no independent existence.

---

[9] Alternatively, one might suggest, as I have above, that the problem of transference is a conceptual problem, which no amount of further empirical evidence could ameliorate. I do not think this can be Locke's position, though, since he clearly regards it as evidence for his claim that we are ignorant of the real essences of bodies. It is also worth noting that Newton's agnosticism about the fundamental causes of observable realities allows him to sidestep the problem of impulse; see Stein (2002: 276 ff.).   [10] Downing (1998: 410).

[11] It is worth recalling that the late scholastics had also used the language of 'primary' and 'secondary' qualities, where the former were singled out by their prominent role in physical explanations. I suspect that this use carries over into Locke; thus, the core notion of a primary quality as that which is fundamental in physical changes is retained, while the qualities thought to fall under this heading, obviously, changed drastically. See M. A. Stewart, introduction to Boyle (1991).   [12] Letter to Stillingfleet (Locke 1823: iv. 467–8; my emphasis).

[13] For a different reading, see Stein (1990: 33). On Stein's view, Locke is saying that Newton convinced him that matter has powers that are miraculous and swing free of any other features of matter. I think Stein misses the epistemic point Locke is making.

We can confirm this by turning to our next trouble spot, ideas of secondary qualities. In perhaps the clearest passage in which Locke discusses God's contribution to the natural world, he speaks with undisguised scorn. It is true that we cannot conceive how the motion of insensible particles can give rise to the ideas of color or sound, since 'Motion, *according to the utmost reach of our Ideas*' (my emphasis) can give rise to nothing but motion. Given this, 'we are fain to quit our Reason, go beyond our *Ideas*, and attribute it wholly to the good Pleasure of our Maker' (IV. iii. 6: 541). To 'quit one's Reason' and ascribe this power to the arbitrary will of the Maker is not to offer a plausible hypothesis but to throw up one's hands.[14]

Although Locke presents the problem of secondary qualities alongside those of cohesion and impulse, this is deeply misleading. For the problem of secondary qualities is not in the first instance a problem to be solved by a science of the natural world. Recall that powers are polyadic properties. To say that bodies have the power to produce sensations of yellow in us is to say something about the bodies but also about our minds in which these ideas exist. And if we are ignorant of the real essences of bodies, 'we are much more in the dark in reference to Spirits' (IV. iii. 17: 548; see II. xiii). Solving this problem would require, but would not be achieved by, knowing the micro-level primary qualities of bodies. That would still leave us with the explanatory gap contemporary philosophers of mind typically point to: why should it be that this or that physical state gives rise to this or that quale? Answering *that* question is a job for the philosopher or scientist of the mind, not the natural philosopher.

Thus, if our question is, are there powers that fail to supervene on the intrinsic properties of the relata, Locke must answer in the negative. Otherwise, relations would have to be real in a way inconsistent with Locke's official view. If all relations are reducible, there is just no way for God or any one else to add powers to things over and above their intrinsic properties, for the simple reason that there *can be* nothing over and above these properties and the things that possess them. Whatever story we tell about Locke's mechanism must square with his conceptual foundationalism.

It is unfortunate that most commentators run together three positions I think we need to keep distinct: course-of-nature mechanism, ontological mechanism, and corpuscularianism. The third is one way to meld the other two: it is one story about how both kinds of mechanism could be true. Locke has 'instanced the corpuscularian Hypothesis, as that which is thought to go farthest in an intelligible Explication of the Qualities of Bodies' (IV. iii. 16: 547), though he doubts that we will be able to find a superior such hypothesis. This kind of hesitancy does not infect his commitment to course-of-nature mechanism, which

---

[14] *Pace* Downing, who seems to think that superaddition in this sense is, for Locke, 'a plausible hypothesis' (1998: 409).

in Locke's case amounts to the claim that bodies act solely by virtue of their primary qualities. ''Tis evident,' Locke writes, 'that the bulk, figure, and motion of several Bodies about us, produce in us several sensations' (IV. iii. 28: 558) and, presumably, produce changes in other bodies as well. This is beyond question. Since bodies act only by motion, the qualities by which they act must be relevant to the course of that motion, and only primary qualities (Q1s and Q2s) can do *that*.

If we thus distinguish mechanism in both forms from corpuscularianism, the puzzling passages of IV. iii snap into place. Locke's doubts concern only corpuscularianism. Locke never backs off from his claim that the powers of bodies consist 'in a Texture and Motion of Parts' (IV. iii. 16: 547). Corpuscularianism posits atom-like entities as these parts; but it is not in principle the only game in town.

To see the difference between corpuscularianism as a defeasible hypothesis on one hand and course-of-nature and ontological mechanism as bedrock commitments on the other, consider Locke's remarks about secondary qualities. He hypothesizes that these secondary qualities depend 'upon the primary Qualities of their minute and insensible parts; or if not upon them, upon something yet more remote from our Comprehension' (IV. iii. 11: 544). Locke is questioning whether it is the minute parts of bodies whose primary qualities explain their secondary ones, not whether it is their primary qualities *tout court*.

The deficiencies of corpuscularianism, then, are not to be held against course-of-nature or ontological mechanism. They are problems that arise within the context of a particular proposal that meets the requirements set by both mechanisms. Whether an idealized epistemic position would remedy them by revealing to us the precise, purely mechanical workings of corpuscles or would instead reveal fundamental elements other than corpuscles is, as it must be, an open question. Boyle himself, in his 'Cosmical Suspicions,' broached the possibility of 'some other kind of Corpuscles' beyond those posited by the 'new Philosophers' in their treatment of the ether (CS 303).[15] Either possibility would preserve the strong bottom-up nature of Locke's position.

To draw this out, consider Newton's own position on gravity. Surprisingly, Newton rejects the possibility of unmediated action at a distance. In his famous letter to Bentley of 1692/3, Newton writes,

It is inconceivable that inanimate, brute matter should, without the mediation of something else, which is not material, operate upon and affect other matter without mutual contact, as it must be, if gravitation in the sense of Epicurus, be essential and inherent in it. And this is one reason why I desired you would not ascribe innate gravity to me. That gravity should be innate, inherent, and essential to matter, so that one body may act upon another at a distance through a vacuum without the mediation of anything else, by and through which their action and force may be conveyed from one to another,

---

[15] I thank Jacob Adler for pointing this passage out to me.

is to me so great an absurdity, that I believe no man who has in philosophical matters a competent faculty of thinking can ever fall into it.[16]

Locke might agree, in so far as he suggests there is a 'fluid medium' connecting all bodies. But Newton's point is not just that unmediated action at a distance is impossible; he thinks, just as the voluntarists have it, that 'Gravity must be caused by an agent acting constantly according to certain laws.'[17] Thus, what Newton rejects is not merely unmediated action but action that is not itself directed by an agent who acts according to laws. This could not be further from the hypothesis that Locke admits; again, the phenomenon of gravitation proves that God has '*put into bodies* powers and ways of operation, above what can be derived from our idea of body.'[18] It does not prove that there is an agent directing the behavior of bodies according to laws. So far from accommodating Newton's view, as Locke seems to take himself to be doing, he in fact endorses a position Newton regards as an absurdity. From Newton's point of view, Locke evidently does not have a 'competent faculty of thinking.'

This position is made more explicit in *Some Thoughts Concerning Education*, where Locke argues that gravity is a result of matter being arranged by God. The relevant passages occur in a totally different context from that of the *Essay*, and we need to be aware of just how different Locke's purposes in the former text are. Locke is arguing that the study of Scripture be undertaken before one studies natural philosophy, to counter the impression that natural philosophy is a science, whereas these other two areas of inquiry are not. Locke thinks it vital that the student not be misled by the apparent perspicuity of the study of bodies, and the consequent unflattering appearance of all other areas of inquiry, and thus be drawn to atheistic materialism. Again, as in the *Essay*, Locke is at pains to emphasize that the ultimate natures of minds and bodies are equally obscure. Locke writes,

it is evident, that by mere matter and motion, none of the great phaenomena of nature can be resolved; to instance but in that common one of gravity, which I think impossible to be explained by any natural operation of matter, or any other law of motion, but the positive will of a superior Being so ordering it. (§19)[19]

What does Locke mean by 'any natural operation of matter, or any other law of motion'? The point is that matter, left to its own devices, even assuming motion, would not so arrange itself as to exhibit the phenomenon of gravity, just as matter without motion would not, by 'any natural operation,' move.[20] This is confirmed by the last part of the sentence: what God does is 'order' (i.e., arrange) matter so that it behaves in a certain way.

---

[16] In Newton (2004: 102).   [17] (2004: 103).
[18] Letter to Stillingfleet (Locke 1823: iv. 467–8; my emphasis).
[19] (1823: ix. 184). Matthew Stuart (1998) makes much of this passage, in the service of Leibniz's interpretation of Locke. For a reply to Stuart, see Downing (2007: 376).
[20] See the Second Reply to Stillingfleet (Locke 1823: iv. 461).

To support this, let's see how Locke continues in this passage. Recall that Locke is arguing that a study of Scripture is a necessary precursor to the study of natural philosophy. For there are many things in Scripture that cannot be explained by 'mere matter and motion.' This does not mean, however, that matter and motion, plus the initial activity of God, is not a sufficient explanation. It looks at first glance, in other words, as if Locke ought to be arguing that the natural world cannot be exhaustively characterized by the intrinsic, mechanically acceptable qualities of bodies. On the contrary, he is really arguing that the natural world cannot be *explained* solely on the basis of these qualities; one needs to add the activity of a divine being. That activity, in the case of gravity, amounts to God 'ordering' the bits of matter in a certain way. In the biblical miracles, God just orders them a bit differently for a time. Locke continues,

And therefore since the deluge cannot be well explained, without admitting something out of the ordinary course of nature, I propose it to be considered, whether God's altering the center of gravity in the earth for a time, (a thing as intelligible as gravity itself, which perhaps a little variation of causes, unknown to us, would produce), will not more easily account for Noah's flood, than any hypothesis yet made use of, to solve it.[21]

The parenthetical remark suggests that gravity itself is the result of unknown causes in the bodies themselves.[22] It is by fooling about with the underlying cause in bodies that God brings about his miraculous effect; but to bring about gravity itself, God must do the same thing, though not on a perpetual basis.

Recall Locke's claim that *none* of the great phenomena of nature can be explained just by matter and motion. In a way, there is nothing special about gravity: it is similarly impossible that peach trees and roses, with their beauty and powers of vegetation, could be the result solely of matter and motion.[23] The doubts Locke expresses here are not about mechanism, whether ontological or course-of-nature, but about the (implicitly atheistic) idea that, left to its own devices, matter, or matter and motion together, could produce even the most common phenomenon of nature.

To sum up: the 'pessimistic' passages have, on my reading, three intertwined goals. First, they show that we do not now have a demonstrative science of bodies. This is one point of his arguments against the pretensions of corpuscularianism to *scientia*. Second, they show that matter on its own, plus or minus motion,

---

[21] Second Reply to Stillingfleet (Locke 1823: iv. 461).

[22] I owe this point to Downing (2007: 376).

[23] As Locke writes in a letter to Stillingfleet: 'The idea of matter is an extended solid substance, wherever there is such a substance, there is matter, and the essence of matter, whatever other qualities, not contained in that essence, it shall please God to superadd to it. For example, God creates an extended solid substance, without superadding anything else to it, and so we may consider it at rest: to some parts of it he superadds motion, but it still has the essence of matter: other parts of it he frames into plants, with all the excellencies of vegetation, life, and beauty, which are to be found in a rose or a peach-tree, &c. above the essence of matter in general, but it is still matter: to other parts he adds sense and spontaneous motion, and those other properties that are to be found in an elephant' (1823: iv. 364).

is not a sufficient and complete explanation of many phenomena. Note that the first point holds only in practice; there is no suggestion that in principle such a science is impossible. The second, however, is not epistemic: the claim is that, even if the secrets of nature were revealed to us, we would still need to posit a God to set the whole business up in the right way. Finally, the pessimistic passages try to show that the conceptual apparatus required by a good philosophy of mind is no more or less mired in confusion than that required by a good philosophy of body. The appearance of inconsistency between these passages and the 'optimistic' passages disappears once one recognizes that nothing in the former threatens the (in principle) possibility of demonstrative science in the latter.

If I am right, Locke emerges as a philosopher with a consistent and quite reasonable view, in its context. He is committed to both forms of mechanism; about the corpuscular hypothesis, he has his doubts. Moreover, his position is the first we have seen that demotes God to the originator of motion; that is, it is the first that opens up the possibility of a completed science that would be all but silent on the nature and activity of the deity.[24]

There remains a serious challenge to my view, however. Edwin McCann has argued that the voluntarist reading of Locke need not violate ontological mechanism. Now, I have argued that the tension Margaret Wilson initially diagnosed between the 'optimistic passages,' which endorse course-of-nature mechanism, and the pessimistic passages, which seem to deny it, is illusory.[25] It would be surprising if some such reconciliation were impossible, given that most of these passages occur within a single chapter of the *Essay* (IV. iii). McCann offers a quite different way of dissolving the tension, one that accords with the voluntarist reading. On McCann's view, God has arbitrarily instituted necessary connections (between, for example, primary qualities in bodies and our phenomenal experience of secondary qualities). This institution is supposed to be enough to ground the in principle possibility of *scientia*. Although the connection 'would not be a "rational" one, since its necessity would not be demonstrable independently of the fact that God had expressly (and arbitrarily) ordained such a connection,' we could construct a strict demonstration on its basis, if only God told us about it through revelation.[26] McCann offers an ingenious means of reconciling this arbitrary connection with ontological mechanism. Rather than constituting an addition to the mechanical properties of objects, McCann says that the powers accruing to objects in virtue of this arbitrary connection 'are not due to any real, nonmechanical component or constituent of the body; the only causally active qualities of the body are its mechanical affections. It is just that, given the laws God has established, these affections are capable of producing

---

[24] With the exception of miracles like the Flood, of course.
[25] See Wilson (1979).   [26] McCann (1994: 72).

the effects in question.'[27] So McCann's voluntarism seemingly does not violate ontological mechanism, as Leibniz feared. If McCann is right, my attempt to reject voluntarism in favor of course-of-nature mechanism must fail.

The first problem with McCann's view is the strained reading he must give of the optimistic passages, such as IV. iii. 25: 556. As we have seen, Locke there claims that we could 'know without Trial' the operations of bodies on one another if we knew 'the Figure, Size, Texture, and Motion' of the microscopic constituents of those bodies. There is no suggestion here, or in any of the other optimistic passages, that one in such an epistemically privileged position would also require a special revelation from God in order to carry out the deduction. But my main difficulty with this proposal lies in understanding the ontology of McCann's 'laws of nature.' Locke, as we have seen, only once uses the phrase outside of the context of ethics, where it has a very different meaning. What could it mean for God to arbitrarily decree laws of nature, in anything like the contemporary sense? Was this sense even available to Locke? However we answer these questions, we must remember that, like the Aristotelians, Locke's primary notion is that of power, not law.[28]

But we can do better than this as a reply. We know that on the one occasion when Locke speaks of laws of nature in the extra-moral sense in the *Essay*, he must be doing so metaphorically. A (literal) Lockean law of nature requires not just a lawgiver but also the threat of punishment and the promise of reward (II. xxviii. 5–6: 351–2). In the sole passage that uses 'law' in connection with bodies (IV. iii. 29), Locke claims that, when we observe regularities, we can be sure that the bodies involved 'act by a Law set them; but yet by a Law, that we know not' (IV. iii. 29: 560). Now, as we have seen, Locke does not think bodies act at all; strictly speaking, they merely transmit motion. Moreover, he can hardly believe that bodies stand under the threat of punishment. His talk of laws is clearly metaphorical, then; but how to cash it out? Locke's text does not tell us, and given the isolated and indeed unique nature of his remark about laws, I am not inclined to offer one. If one must be had, however, it is ready to hand: Locke might well simply mean 'law' in Boyle's sense, that is, as a regular and steady course of nature. This would square with his emphasis on powers and allow us to make sense of his epistemic pessimism. While we can know that events take the course they do for some ends or others, and in regular patterns, the explanation of these patterns, so long as we remain ignorant of the constitution of the bodies around us, cannot be known.

---

[27] McCann (1994: 75).

[28] Something like McCann's view, I have argued, can be sensibly applied to Boyle. But that is because Boyle's God is constantly supplying and regulating motion. I find no indication of a similar view in Locke; indeed, the talk of bodies as possessing passive powers simply rules out such an interpretation.

22

Conclusion

Any attempt to resuscitate the powers of created beings, and hence the bottom-up picture, must navigate between two obstacles. On one side lies occultism: given ontological mechanism, there seems no place for powers in nature, and nothing for them to do even if there were. On the other lurks voluntarism: if God is directly responsible for the distribution of motion, the bottom-up view has been sacrificed, and the supervenience base of the powers of bodies must include more than their intrinsic properties.

We can now see precisely how Locke threads his way between these two pitfalls. He avoids the mysterious ontology of the scholastics by analyzing powers in much the same way as Boyle, though he goes much further and endorses the robust claim. And he avoids voluntarism by parting ways with Boyle over the status of the laws of motion and hence God's role in supplying and regulating motion itself. Once motion is introduced in the world, causal processes take the course they do in virtue of the primary qualities (Q1s and Q2s) of objects. If I am right, the reducibility of powers is a central feature of anti-Cartesian mechanism, and provides a way of anchoring the course of nature in the natures of things themselves.

PART IV
HUME

INTRODUCTION

Everyone we have considered so far, even Malebranche and Berkeley, has been a realist about causation. The questions have been, what kinds of things are causes, and how do they operate? In Hume, we find our first thoroughgoing anti-realist: there is and can be no such thing as power, or cause, as construed by these realist philosophers. This reading is controversial. Defending it will lead us into the heart of the innovations Hume effects that make his conclusions inevitable. Although many of Hume's weapons were forged by his intellectual progenitors—the theory of relations, the argument from nonsense, and the 'no necessary connection' argument, most importantly—Hume recasts them, and turns them against their realist inventors.

23

The Two Humes

Reading his texts from a contemporary perspective, there can easily seem to be two Humes. One is the Hume of the subtitle of the *Treatise*: 'being an attempt to introduce the experimental method of reasoning into moral subjects.' The chief concern of this Hume is to map the cognitive structures of the mind. By tracing, although not ultimately explaining, the cognitive mechanisms that govern relations between perceptions (ideas and impressions), this Hume seeks to emulate his hero Newton. This Hume is uninterested in metaphysics as such; his concern is with how we as a matter of fact come to think of the world and ourselves.[1]

The other Hume, by contrast, does not hesitate to draw metaphysical conclusions: there *is* no such thing as a Cartesian self, an external world, or causation as realists conceive of it, Causation with a capital 'C,' to follow Galen Strawson.[2]

Assuming these two projects are in tension, many commentators seize on one. Those persuaded of a naturalist reading of Hume emphasize the psychological project at the expense of all else. That is, they read Hume as primarily interested in connections among perceptions and at most agnostic on the nature of things.[3] No doubt those interested in metaphysics and epistemology have been equally guilty of playing up the other side of Hume, which is unabashedly concerned with topics all contemporary philosophers recognize as their particular province, rather than that of psychologists.

---

[1] This is the side of Hume emphasized by Norman Kemp Smith, who quotes Newton: 'These Principles [such as Inertia, Gravity, Cohesion of Bodies], I consider, not as occult Qualities, supposed to result from the specifick Forms of Things, but as general Laws of Nature, by which the Things themselves are form'd; their Truth appearing to us by Phaenomena, though their Causes be not yet discover'd. For these are manifest Qualities, and their Causes only are occult' (Kemp Smith 1941: 55). Similarly, Kemp Smith suggests, Hume can be agnostic about the ultimate causes of the principles of mental attraction he picks out. Another, more recent writer who seizes on this side of Hume is Helen Beebee (2006), who defends a version of the 'New Hume' reading, discussed below, as does P. J. E. Kail (2007).

[2] I shall reserve 'Causation' for this realist view, while 'causation' will serve as a theory-neutral term.

[3] See e.g. Garrett (1997) and perhaps Owen (1999). Despite appearances, the reading I shall defend is not so distant from Garrett's. Garrett argues that, although Hume's conclusion directly concerns the psychological mechanisms responsible for induction, it 'nevertheless provides good reason to conclude that no argument can show the reliability of induction by argument without *presupposing* that reliability' (1997: 94).

Both Humes are really there in the text. But the opposing readings to which they give rise are inconsistent, and only one is warranted. Kenneth Winkler has helpfully dubbed the reading that takes Hume to be a realist about Causation the 'new' Hume, as opposed to the 'old' Hume, who does not confine himself to the psychological project and declares that there is no such thing as Causation.[4] I shall argue in Chapter 24 below that Hume's account of intentionality precludes his recognition of even the in principle possibility of Causation, that is, causation understood as something more in the nature of things than constant conjunction.

Before we can sort this out, however, we have to begin by seeing how the two strands of Hume's thought are woven together. Both are well stated in the introduction to the *Treatise*. As is usual among British empiricists (and Kant as well), Hume begins by decrying the sorry state of philosophy and human knowledge generally. His solution is, not surprisingly, a version of Locke's own 'Historical, plain' method:

> Here then is the only expedient, from which we can hope for success in our philosophical researches, to leave the tedious lingring method, which we have hitherto follow'd, and instead of taking now and then a castle or village on the frontier, to march up directly to the capital or center of these sciences, to human nature itself; which being once masters of, we may everywhere else hope for an easy victory. From this station we may extend our conquests over all those sciences, which more intimately concern human life, and may afterwards proceed at leisure to discover more fully those, which are the objects of pure curiosity. There is no question of importance, whose decision is not compriz'd in the science of man; and there is none, which can be decided with any certainty, before we become acquainted with that science. In pretending therefore to explain the principles of human nature, we in effect propose a compleat system of the sciences, built on a foundation almost entirely new, and the only one upon which they can stand with any security.   (*T* I. 6)

Explaining the principles of human nature does not, for Hume, require offering an explanation that would transcend experience; as we shall see, no such explanation is intelligible. All Hume can do, and needs to do, is explain the phenomena in the sense of appealing to the most general principles or rules by which the mind moves from one perception to another (*T* I. 8). Although often abused for not having understood Newton, Hume here follows the Newton of the Queries to the *Opticks*, and that in two ways: first, in speaking of qualities or forces as principles or rules (a usage that does not come naturally today),[5] and second, in remaining agnostic on the ultimate explanation of these principles.

---

[4] See Winkler (1991). Winkler himself defends an agnostic reading of Hume, which I discuss below. Strictly speaking, Winkler's classification applies only to opposed readings of Hume on causation. But it is easy to see how it can be generalized to cover Hume's approach to metaphysics and epistemology generally.

[5] See Newton (2004: 137). In the *Treatise*, Hume does not, as far as I can tell, use the expression 'law of nature,' except in the moral sense, though in the *Enquiry Concerning Human Understanding* he uses it quite frequently to mean simply a well-confirmed regularity.

To my mind, there are three levels to Hume's project. The first is what we might call the ambition to be the Newton of the mind, to show how in fact we make transitions from one perception to another. The second is to show that any hope of a justification for such inferences is illusory. The bottom level makes good on the goal of the top, by revealing imagination or habit as the basic faculty, the 'gentle force' that explains the workings of the mind.

Hume's fundamental project, then, is the psychological one of identifying explanations, whether or not they are also *justifications*, for the beliefs he thinks the mind is disposed to form. But as Hume indicates above, this project, followed to its conclusion, is supposed to have consequences for every 'question of importance' in philosophy and the sciences. How can this be? For philosophers trained to make a sharp distinction between the normative and the descriptive, this latter ambition seems like a category mistake. Nothing about how beliefs happen to be formed will tell you what is in fact the case.

Here is where the middle layer of Hume's project comes into play. Developing his overall strategy, Hume identifies four possible sources for the transitions among perceptions he wants to explain: demonstrative reasoning, as one finds in mathematics,[6] non-demonstrative reasoning based on experience,[7] sensation (i.e., impressions themselves), and imagination or fancy (the 'gentle force'). In most cases, and in the most philosophically interesting cases, Hume eliminates the first three of these as possible sources of our beliefs, and leaves imagination in possession of the field. Moreover, once we have a grip on Hume's theory of relations, we shall see that there is no difference in kind between the two forms of reasoning; reason is nothing but a 'wonderful and unintelligible instinct' (*T* 1. 3. 16. 9). Another way to put this point, which some will prefer, is to say that, rather than rejecting demonstrative and non-demonstrative inferences as sources of our beliefs, Hume transforms them by showing how they depend upon the imagination.

In the case of causation, for example, Hume begins by testing the hypothesis that our inference from cause to effect, that is, the habitual motion of the mind from an impression of a cause to the idea of its effect, is justified by demonstrative reasoning. Such a justification would have to depend on the idea of a necessary connection between the two. Hume concludes that this gets matters exactly backwards. The necessary connection depends on the inference, not the other way round (*T* 1. 3. 14. 21). Differently put, it is the customary transition from one perception to the other that generates our impression, and hence our idea, of necessity. If such transitions were justified, they would require an impression of necessary connection independent of this transition. And this is an impression that we simply do not have.

---

[6] Geometry is a special case, since Hume (in the *Treatise*, though not in the *Enquiry*) does not believe it attains the same level of certainty as pure mathematics.

[7] As Garrett (1997: 94) shows, Hume uses 'reason' univocally to refer to the first two categories I list here.

Thus, the two Humes come together. The properly philosophical conclusions Hume reaches are in the service of his larger psychologistic project. But this does not rob them of their force or interest. No one pursuing the psychological project can neglect the possibility that our inferences are the result of properly functioning rational structures, which provide justifications for our beliefs. And if such a person concludes that that is not in fact the case, he will cast about for a suitable non-rational principle. The implications for allegedly normative areas of inquiry are obvious: swinging entirely free of actual practices and phenomena, they are at best irrelevant; at worst, they are fictions designed to comfort us in the face of the irrationality, or, better, a-rationality, of our belief-forming mechanisms.

All of this still leaves us with the question of Hume's metaphysics. Is Hume a realist about Causation? Is he agnostic, leaving the question undecided? Or does he instead reject Causation full stop? Perhaps one could follow out a purely psychological project that would remain innocent of any commitments or implications, one way or the other, for metaphysics.[8] But this is not Hume's project. He intends his descriptive project to decide the core questions that the Introduction flags.

Settling this requires that we see just *how* reason and sense fail as justifiers of many of our beliefs. In the central case of causation, as well as substance and the external world, these faculties cannot be invoked, because if they were, we would have to have an idea that we cannot have. Some philosophers have, understandably, called this Hume's 'positivism.' I think this is not right, and below I show just how different Hume's approach to meaning is from that of the positivists. But the failure of the positivist reading offers no comfort to those who would take Hume to be a realist or agnostic about Causation. If a word fails to signify an idea, it is simply nonsense. And there is no way to be so much as agnostic about a piece of nonsense.

Thus, the following chapter (24) is devoted to exploring Hume's views on meaning and intentionality. These form the core of his negative project, embodied by the middle layer of his general strategy, since Hume is here concerned to deny that reason or sensation could produce an idea of Causation, or necessity. His general argument at this level is a version, then, of the argument from nonsense: Causation is to be rejected because it is, one way or another, unintelligible. We shall see how Hume's novel conception of the proposition, and the mental act by which it is generated, closes off the possibility of a Lockean definite description account of the meaning of 'substance,' 'cause,' or 'power.'

This restrictive meaning empiricism, however, is only the first step in making good the argument from nonsense. As we shall see, many arguments that will by

---

[8] This is in a sense trivially false, since any psychology will involve ontological commitments to such things as mental states, perceptions, or replacements for these. The real issue, of course, is whether one could pursue such a project without invoking any extra-mental entities.

now be familiar are recast by Hume and take on a decidedly new aspect because they are deployed in the service of the argument from nonsense. Malebranche's epistemic and 'no necessary connection' arguments now have as their conclusion, not that there is no such thing as Causation between finite objects or minds, but that we have no *idea* of such a thing. Chapter 25 canvasses these arguments as Hume uses them, and exposes his real objections to the geometrical and cognitive models of causation we have been considering.

Alongside the argument from nonsense, Hume, at this middle level of his strategy, also sets up what I call 'the practicality requirement': causal reasoning is something we, and non-human animals, engage in all the time, even before we have come to the full use of our faculties. This in effect sets up another constraint on plausible accounts of such reasoning: it must be something not only a child but a bird or orang utan can engage in. The practicality requirement helps give Hume the clue that sets him on the path to his own anti-realist conception, for it is only the brute faculty of the imagination in its most elemental form that can explain the inductive skills of humans and non-humans alike.

After following Hume as he 'beats about all the neighbouring fields,' we will be well positioned to see how he begins to craft his positive view. The final two chapters below (26 and 27), then, are devoted to the third tier of Hume's overall project, his positive account. Before we can grasp Hume's definitions of cause, we need to see what he thinks of relations generally. I argue in Chapter 26 that Hume takes up Locke's conceptual foundationalism and radically modifies it. This interpretation allows us to dissolve a key tension in Hume's view, as it is usually understood: he seems to think both that no perceptions are really related, that is, are related apart from the mind's activity of moving from one to another, and that we 'discover' some relations, which indicates just the opposite. His modified conceptual foundationalism is well suited to reconcile these claims.

Just as Hume splits Locke's ideas into ideas and impressions, so he splits Locke's relations into philosophical and natural relations. In line with some recent commentators, I argue that with regard to those relations that are both natural and philosophical, the natural are prior to and explanatory of the philosophical. Once we see just how different these two senses of 'relation' are, we will be able to see how Hume can account for our ordinary ways of talking about causes. Equally importantly, we shall be able to see that a well-known problem with his two definitions—namely, that they are not coextensive—is not in fact a problem at all, but just what his theory of relations requires.

Ultimately, then, Hume will emerge as a figure whose methods are, in some ways, very much at odds with those of the other philosophers we have looked at. But seeing how his conceptions of intentionality, meaning, and relation, conspire to form a coherent whole will let us see just how the seeds of his view were planted by his antagonists. The old saw that Hume is occasionalism minus God is not too far off the mark. Consider Berkeley's claim that visual ideas are a divine language, provided by the creator to help us navigate through the world. Berkeley, though not a thoroughgoing occasionalist, transforms Malebranche's notion of occasional or natural causes into his own notion of words in the language of God. While not sharing any genuine causal tie, the regularity of visual perceptions is secured by God's benevolence and, one might say, the syntactic rules of visual grammar. With Hume, even this weak connection is gone. The divine language has become gibberish.

24

Intentionality

It is impossible for us to *think* of any thing, which we have not antecedently felt . . .

(E 7. 4)

The argument from nonsense is Hume's main weapon; even the arguments he borrows from Malebranche are transformed so that their conclusions fit this technique of undermining Causal realism by denying its conceivability. Thus, we have to begin investigating the second level of Hume's project by exploring the constraints he places on meaning and thought.

According to devotees of 'the New Hume,' Hume in fact is a Causal realist. The arguments I develop below show that Hume has no means (or wish) to acknowledge even the bare possibility of Causation. This is because the notion does not so much as make sense. *Pace* such writers as Galen Strawson and Janet Broughton, Hume does not believe that, as the latter puts it, we can conjure up 'the *bare thought* . . . of there being some feature of objects that underlies . . . constant conjunctions.'[1]

In criticizing the New Humeans, Kenneth Winkler has carved out a second option, according to which Hume refrains from affirming that there are Causes, while not denying that there are.[2] But Winkler also argues, as I shall below (though in a different way), that Hume holds we cannot '*in any way*' conceive of Causation.[3] Why, then, should Hume hold back from denying that there are Causes? Not being able to conceive of them, he must obviously refrain from affirming their existence. But the only sense in which, if Winkler and I are right, Hume can be said to refuse to deny them is that in which I do not deny that *fongs* exist. That is, if a word is truly nonsensical, it is equally silly to assert or deny propositions involving it. This is not because there might *really, after all* be fongs; it is because there is nothing there to argue about until we specify the meaning of 'fong.'

---

[1] Broughton (1987: 235), also quoted in Winkler (1991). Kail (2007) makes Broughton's notion of the 'Bare Thought' central to his view.   [2] See Winkler (1991: 543).

[3] Winkler (1991: 576).

Wade Robison has rightly pointed out that denying meaning to a claim is not the same as denying that claim.[4] There is no comfort here, though, for any reading of Hume as leaving the status of causal or ontological claims up in the air, *pace* Robison. A meaningless claim does not even rise to the level of having a truth value, one way or the other; it would be a strange trick to turn this into the very means by which that claim rises above rational evaluation. There is no room for a position that denies all meaning to Causation and at the same time remains agnostic, however tentatively and anemically, on its existence.[5]

A helpful way to frame this debate on intentionality and its limits is to contrast Hume with Locke. As we have seen, Locke has no problem with extending that which we can sensibly talk about beyond the limits of possible experience. Locke happily speaks of in principle unperceivable entities such as substance,[6] real essence, and so on, even though he denies that there is much we can informatively say about such things. The New Hume reading, whether knowingly or not, takes Hume to be conforming closely to Locke's view. The following sections will show just where this is right and where, crucially, it goes wrong. Hume's meaning empiricism is significantly more restrictive than Locke's.

There are two key elements to Hume's meaning empiricism, one at the linguistic level, and another at the level of the mental. First, as we shall see, Hume takes for granted that all words signify perceptions. This demand for an ideational backing, as it were, to any significant word has its source in Descartes's way of ideas, and, as we have seen, Malebranche's own 'argument from nonsense,' as well as Locke's deployment of his linguistic thesis (*Essay*, III. i–iii). Now, this slogan—all words stand for ideas—is not so anodyne or commonplace as it sounds, once we see precisely what it means in Hume's mouth. An object of belief, for Hume, is just an idea, not a complex of acts and ideas, as Locke thought. By eliminating syncategoremata as making any real contribution to the content of thought, Hume also eliminates the possibility of giving words meaning by using definite descriptions. Even a complex idea cannot do the work of a definite description, since it does not include syncategorematic acts. Second, at the level of the mental, Hume claims that all ideas are copies of impressions. This 'copy principle' maintains, more precisely, that the only intrinsic differences between an idea and an impression are their force, vivacity, or liveliness. Any difference that affected what we might call the content (or intrinsic nature) of either the idea or the impression would prove that one was not in fact the copy of the other. Hume's restrictive meaning empiricism consists in the combination of these two claims. Together, they rule out Causation as even an in principle possibility.

[4] See Robison (1976). Robison is an early defender of the 'New Hume,' though he focuses on the external world rather than causation.
[5] But see the discussion of philosophical as opposed to ordinary nonsense below.
[6] That is, substance in the pincushion sense, not in the sense of substances such as trees and dogs.

24.1 MEANING

Despite all the attention recently paid to the role of Hume's theory of meaning in his attack on causal realism, very little has been done to elucidate that theory. Many commentators simply ignore Hume's claims or presuppositions about language, diving right into his theory of ideas. Thus, Jonathan Bennett's section on Hume's 'meaning empiricism' focuses on the principle that all ideas are copies of impressions, which is not itself a claim about meaning.[7] Barry Stroud's and Don Garrett's impressive books simply do not discuss his views on language at all.[8] Perhaps some commentators are assuming that Hume's view is at once transparently clear and transparently borrowed from Locke, as if citing a pedigree were an elucidation.[9] When commentators do address the issue, however briefly, their claims are invariably off the mark. Hume has been called a 'verificationist' (Rosenberg); he has been said to hold that words refer to ideas (Beauchamp and Rosenberg; Stanford).[10] None of these accurately reflects Hume's position.

A typical response to the Old Hume reading I am about to offer is that it is underpinned by a projection of twentieth-century positivism into Hume's text.[11] After setting out the bare outlines of Hume's theory, I explain precisely how it differs from positivism. At the same time, as we shall see, this gives no comfort to the notion that Hume might allow for unexperienceable Causal connections.

Like previous empiricists, Hume takes for granted that words signify ideas. And he takes all ideas to be copies of impressions. Together, these points constitute a powerful instrument with which to examine the claims of other philosophers (*E* 7. 4). In the Abstract, for example, Hume writes: 'when [the author] suspects that any philosophical term has no idea annexed to it (as is too common) he always asks, *From what impression is that pretended idea derived?* And if no impression can be produced, he concludes that the term is altogether insignificant' (*T* 409). And in the Appendix, Hume claims that if we wish to speak of 'self' or 'substance,' 'we must have an idea annex'd to these terms,

[7] See Bennett (1971). I discuss Bennett's position below. I do not mean to imply that the copy principle does not have *implications* for Hume's theory of meaning; thus, I discuss appeals to this principle in the context of verificationism.

[8] See Stroud (1977) and Garrett (1997). Those suspicious of my claim need only glance at the indexes of these books: neither has entries for 'language,' 'meaning,' 'words,' 'reference,' etc.

[9] See e.g. Rosenberg (1993). Rosenberg takes Hume to be a positivist who adopts Locke's views on language. (I should acknowledge here that I am indebted to Rosenberg for his careful and astute comments on this material.) Writers such as Simon Blackburn (1993) and John P. Wright (1983) discuss Hume's views in the context of positivism, without explaining precisely what they take Hume's position on the connection between words and ideas to be, or how 'positivism' is to be understood in a Humean context. Galen Strawson (1989) has much to say about Hume on meaning but once again ends up focusing on his theory of ideas. More examples could be cited.

[10] Beauchamp and Rosenberg (1981); Stanford (2002).

[11] See e.g. Livingston (1984: 150) and Strawson (1989: 179), quoted in Winkler (1991).

otherwise they are altogether unintelligible' (*T* 399). Similar pronouncements can be found throughout Hume's writings.[12]

In such passages, we find Hume mentioning two relations, 'annex' and 'signify.' These turn out to be one: an idea is said to be 'annex't' to a word, and a word 'signifies' an idea (or impression). The only difference is the direction in which the relation is said to hold: word–idea or idea–word.

With this admittedly bare sketch in place, we can begin clearing the ground for the correct reading of Hume. A useful foil here is the positivist reading, which takes Hume to have anticipated early twentieth-century empiricism. Seeing how it goes wrong will put us in a better position to see precisely how Hume's empiricism becomes as restrictive as it is.

24.2 AGAINST THE POSITIVIST READING

There are two possible sources of support for the verificationist reading of Hume. The first appeals to Hume's copy principle; the second, to his distinction between relations of ideas and matters of fact. Since no commentator I know of explicitly makes these arguments, I shall do my best to construct them. As we shall see, the copy principle in no way implies verificationism; with the second line of argument, matters are more complex.

What I shall call 'classical verificationism' is the doctrine that any meaningful proposition is either (*a*) analytic or (*b*) capable of being confirmed or disconfirmed by some possible experience. In his introduction to *Language, Truth and Logic*, A. J. Ayer weakens (*b*) and allows meaning to any synthetic proposition for which there can be some empirical evidence one way or the other. This does not require that any synthetic proposition admit of being completely confirmed.

For my purposes, there are two important features of classical verificationism:

1. It is forward-looking. It says exactly nothing about the *origin* of the proposition or its component 'ideas'; what is at issue is how, if at all, experience can be relevant to determining its truth value. (This is true even if we widen the scope of (*b*) to include propositions for which there *could have been* evidence.)[13]

2. It takes whole propositions as its object. A classical verificationist will not wonder about a phrase like 'the external world' in isolation; the strategy is

---

[12] e.g. *T* 1. 1. 7. 1; 1. 3. 14. 14; see also *E* 4. More common, however, are arguments that depend on this principle. In describing how we come to believe that Caesar was killed on the Ides of March, Hume describes the inferences we make on the basis of the work of historians: 'Here are certain characters and letters present either to our memory or our senses; which characters we likewise remember to have been us'd as the signs of certain ideas . . .' (*T* 1. 3. 4: 83).

[13] Admittedly, the tag 'forward-looking' is then less felicitous. But the point holds: the verificationist is asking whether evidence can (or could have) accrued to a proposition, not where the constituent ideas of that proposition came from.

always to take such phrases and put them in sentences that make existential claims. The verificationist looks to see how one might go about determining *whether there is* such a thing as an external world, causation, or the self, not how we represent these things in the first place.

Why might one think Hume is committed to verificationism? Perhaps the principle that every idea is a copy of an impression will suffice. But the copy principle applies to ideas, not words: it is not a thesis about how sentences or words get their meaning, but about how ideas and impressions are related. To be sure, it has consequences for a theory of meaning, but it is not itself *about* any of these linguistic issues. In the *Enquiry Concerning Human Understanding*, for example, Hume discusses the copy principle in the context of his attempt to 'fix, if possible, the precise meaning' of terms like '*power, force, energy,* or *necessary connexion*' (*E* 7. 3). There is no way to fix the meanings of terms without specifying the ideas they signify. But there is an extra step here: Hume requires that we 'Produce the impressions or original sentiments, from which the ideas are copied' (*E* 7. 4). Fixing a term's meaning is just settling what idea it signifies. The distinctive element Hume adds to this commonplace demand for clarity is that one trace the idea back to its origins in experience. But again, this functions at the level of ideas, not words. Thus, I think it is a mistake for Jonathan Bennett to take Hume's 'meaning empiricism' to consist in his use of the copy principle.[14] The copy principle cannot by itself commit Hume to verificationism. To do so it would need to be supplemented by a thesis about the relation between ideas and words.[15]

Even if we put this aside, the copy principle has neither of the two features above. Unlike the verification principle, it is essentially backward-looking, or genetic. It asks, from what impression is this idea copied, not, how might one go about finding out whether there is something corresponding to the idea outside the mind. This is closely connected with the second feature: the copy principle, as Hume deploys it, is directed at parts of propositions, not propositions themselves. This is true even though Hume thinks of propositions as nothing but lively ideas.

We can see this more clearly if we look at Hume on the self. Keeping in mind the quotations from the Appendix above, we can see that Hume wonders, not how we might confirm the existence of a Cartesian ego, but rather, what entitles us to speak of such a self in the first place. If we cannot trace the alleged idea back to an impression, there is no idea there at all.[16]

---

[14] Bennett (1971: 225 ff.).

[15] Another way to put this is to say that verificationism is supposed to tell us when sentences have meanings and when they do not; but the copy principle tells us where an idea came from, not what a string of symbols signifies. A sentence gets its meaning by signifying an idea; there is no corresponding further fact in virtue of which an *idea* has meaning. I owe this point to an anonymous referee.

[16] Rosenberg (1993) argues that Hume anticipates, not classical verificationism, but an earlier version of the theory according to which any meaningful term 'require[s] a set of observationally

There is a more plausible way to argue that Hume is a verificationist, however. This approach would appeal, not to the copy principle, but to Hume's famous distinction (in the *Enquiry*) between matters of fact and relations of ideas, two categories of propositions which he claims are both exhaustive and mutually exclusive. (Although Hume does not juxtapose 'relations of ideas' and 'matters of fact' in the *Treatise*, he in effect draws the same distinction.) Any object of human reasoning must fall into one or the other, and these just seem to be categories (*a*) and (*b*) above. Why isn't this enough to make him a verificationist?

First, it is not clear that category (*a*)—analytic statements—maps straight-forwardly onto relations of ideas. This depends, of course, on how one defines analyticity. But suppose we take an analytic sentence to be one whose denial issues in a contradiction, something of the form 'A = A' or reducible to this form by the substitution of synonymous expressions. If this is our definition, Hume's notion of a relation of ideas is at least not cointensive, and perhaps not coextensive, with analyticity. Let us get a bit clearer on just what Hume's distinction amounts to. Some propositions can be known simply by inspecting our ideas, while others can be known only through experience (*T* 1. 3. 1. 1). In the latter cases (what Hume in the *Enquiry* would call 'matters of fact'), the order in which the constituent ideas come to us is relevant to determining the proposition's truth value. If an impression of two billiard balls moving rapidly away from one another always preceded an impression of their collision, we would not judge that the collision caused the movement. By contrast, the relations of resemblance, contrariety, degree in quality, and proportion in quantity or number make no reference to such extrinsic facts. Of these four relations, only the last is the object of demonstration; the remaining three belong to intuition. That yellow is lighter than orange (a difference in degree) can be known simply by looking at the relevant ideas, but it is not analytic on the above definition.[17]

What is more important, there is at least one decisive case where Hume appeals to the distinction between relations of ideas and matters of fact to show, not

necessary and sufficient conditions of application.' According to Rosenberg, Hume's 'semantic claim that ideas refer to impressions' lands him with a view 'indistinguishable' from this early verificationism (1993: 66). First, it is not clear in what sense an idea can be said to *refer* to an impression. As we have seen, ideas can signify other ideas, and impressions can signify ideas. And when someone uses a term without signifying an idea by it, she utters nonsense. But none of this implies that ideas *refer* to impressions. Second, although early verificationism seems to have lacked the whole-proposition feature (2) above, it retains the forward-looking feature (1): the criterion states that there must be a set of conditions under which it would be legitimate to use the term in question. By contrast, Hume's copy principle asks about the origins of the mental item allegedly signified by a given term.

[17] Relations of ideas might map onto some other contemporary readings of analyticity. I do think, however, that there are some sentences, e.g., 'No surface can be red and blue all over at the same time,' that do not fall comfortably into the realm of the analytic on any contemporary understanding of the concept and yet would be regarded by Hume as true in virtue of a relation of ideas.

that a given claim is meaningless, but that there is no justification for it. In the *Treatise*, Hume argues that if reason were responsible for our inference from an impression to an idea, it would have to 'proceed upon that principle ["UN"], *that instances, of which we have had no Experience, must resemble those, of which we have had experience, and that the course of nature continues always uniformly the same*' (1. 3. 6. 4). But how is UN to be supported? Hume divides the possible sources of argument into knowledge and probability. In the *Treatise*, as we have seen, only four relations yield genuine necessity (that is, their denial is inconceivable, either because it is self-contradictory or can be seen to be impossible by inspecting the relevant ideas). Let us call these N-relations. If we could argue to UN by appeal to N-relations, UN itself would have to be an N-relation, that is, it would have to be impossible for it to be false. This is clearly not the case. Since UN is not a necessary truth, it must be supported, if at all, by appeal to experience. But this, of course, would beg the question, since all inferences from experience depend on, and so cannot support, UN. But neither in the *Treatise* nor in the *Enquiry* does Hume suggest that UN is meaningless or nonsensical. UN is perfectly intelligible: there is no difficulty envisioning a state of affairs in which it holds true. The difficulty comes only when we try to justify it. Thus, the matter of fact–relation of ideas distinction functions in at least one case to show that a claim is not meaningless but unjustified. By contrast, the verificationist must claim that UN is gibberish, since it falls into neither class (*a*) nor (*b*).

If I am right, there are some important differences between Hume and the verificationists. His distinction between matters of fact and relations of ideas does not map onto the analytic–synthetic distinction, and empirical unverifiability does not entail meaninglessness.

24.3 SIGNIFICATION

Hume, then, is not a positivist, of any stripe. But that is not to say that his view is sufficiently permissive to allow terms like 'causation' to refer to experience-transcending powers or objects. To get Hume's real view in focus, we need to begin with Locke's.

Unfortunately, little about Locke's views is uncontroversial. I shall present my own interpretation as briefly as possible and without doing much to defend it, since I have done so at length elsewhere.[18]

Locke's position centers around the thesis that 'Words, as they are used by Men, can properly and immediately signify nothing but *Ideas*, that are in the Mind of the Speaker . . .' (III. ii. 4: 406). Words themselves have no content; their only use is to serve as signs of ideas, which, of course, *do* have content in

---

[18] See Ott (2004*b*).

the sense that they can pick out or refer to objects and states of affairs, mental and otherwise. Thus, any verbal proposition gets its meaning by a connection to a mental proposition: it signifies a set of ideas and acts in the mind of the speaker. The crucial interpretive question then becomes, what does Locke mean by 'signify' and its relatives? My proposal has been to take as our starting point Locke's claim that *ideas* are signs and then use this to illuminate his notion of signification generally.[19]

Ideas are 'designed to be the Marks, whereby we are to know, and distinguish Things, which we have to do with' (II. xxx. 2: 372–3). It is because I have a certain sensation, say of the color white, that I am justified in inferring the presence of an object in my environment with that quality. Of course, the quality turns out to be nothing more than a power in the object to produce that very idea; the idea does not resemble, but is a reliable indication of, the quality. Thus, whether they are 'only constant Effects, or else exact Resemblances of something in the things themselves,' such ideas are dependable marks or signs[20] of the objects or qualities of objects that can cause us to have those ideas.

Locke's use of 'sign' is in line with a tradition running from Aristotle and the Stoics to Antoine Arnauld and Thomas Hobbes.[21] A 'sign' or 'mark' in this sense is a symptom, a ground for inferring to the presence of something beyond itself. Thus, when Locke says that 'The use then of Words, is to be sensible Marks of Ideas . . .' (III. ii. 1: 405), I take him to mean that words provide one's hearers (or readers) with a symptom or evidence of the speaker's (or writer's) ideas and mental acts. Unlike ideas, of course, words are conventional, as opposed to natural, signs. Nevertheless, both can be reliable indicators of states of affairs.

Later in book III, Locke relaxes his claim that all words signify ideas, allowing that some words signify mental acts. Syncategoremata like 'is,' 'but,' 'if,' and so on signify, not ideas, but instead 'the *connexion* that the Mind gives to *Ideas, or Propositions, one with another*' (III. vii. 1 471). We need words to signify the mental act that as it were ties two or more ideas together into a single thought. Without these, Locke argues in III. vii, there would be no way to generate propositional content.[22] A proposition cannot consist merely of signs of ideas,

---

[19] '[T]here are two sorts of Signs commonly made use of, *viz. Ideas* and Words' (IV. v. 2: 574).

[20] Locke uses the terms 'mark' and 'sign' interchangeably. For example, immediately after stating that words signify ideas in III. ii, Locke writes, 'That then which Words are the Marks of, are the *Ideas* of the Speaker: Nor can any one apply them, as Marks, immediately to anything else, but the *Ideas*, he himself hath' (III. ii. 2: 405). I of course have other arguments for my reading of signification; see Ott (2004*b*, ch. 1).   [21] See Ott (2004*b*, ch. 1).

[22] 'Besides Words, which are names of *Ideas* in the Mind, there are a great many others that are made use of, to signify the *connexion* that the Mind gives to *Ideas, or Propositions, one with another*. The Mind, in communicating its thought to others, does not only need signs of the *Ideas* it has then before it, but others also, to shew or intimate some particular action of its own, at that time, relating to those *Ideas*. This it does in several ways; as, *Is*, and *Is not*, are the general marks

for then it would be a list rather than something that admitted of a truth value.[23] Thus, on Locke's view, an affirmative verbal proposition consists of signs of at least two ideas joined by the mental act signified by 'is.' Other syncategoremata such as 'but' and 'if' can be used to signify the logical connections we take to obtain between the propositions generated by this mental act and the attitudes we adopt toward these propositions.[24]

This will have to serve as a sufficient background for my argument. To what degree does Hume adopt Locke's position on language? As we have already seen, Hume insists that words signify ideas, and that ideas are 'annex't' to words. But it should be equally clear that these are technical terms, whose meaning depends partly on their context. Let us begin by examining Hume's uses of 'signification.'

Although Hume does not explicitly say how he means this term to be understood, we have some important clues before us. Like Locke, Hume often speaks of signs in non-linguistic contexts. If we assume, as we did with Locke, that he is using the term univocally, we can then apply this non-linguistic notion of signification to the linguistic case. Now, Hume holds that a passion that produces sympathy in an observer 'is at first known only by its effects, and by those external signs in the countenance and conversation, which convey an idea of it' (2. 1. 11. 3). More clearly: 'Whether a person openly abuses me, or slyly intimates his contempt, in neither case do I immediately perceive his sentiment or opinion; and 'tis only by signs, that is, by its effects, I become sensible of it' (1. 3. 13. 14; see 1. 3. 16. 6). Putting these passages together, we find the claim that facial expressions, for example, can serve as signs of mental attitudes because they are typically *caused by* a given mental state such as contempt. It seems clear that 'sign' is intended in these passages in Locke's sense: as a symptom or grounds for inference from the sign itself to something unobserved. A blush can serve as a sign of embarrassment, that is, as a reliable indicator of that mental state. My proposal, then, is that Hume construes the workings of language on analogy with these non-linguistic cases of signification, and uses 'sign' univocally in both contexts. After all, in both contexts the chief issue is how we reveal our minds to others. What is needed is an indication or symptom of that which is otherwise hidden. The Lockean notion of a sign is defined precisely by its ability to play

of the Mind, affirming or denying. But besides affirmation, or negation, without which, there is in Words no Truth or Falsehood, the Mind does, in declaring its Sentiments to others, connect, not only the parts of Propositions, but whole Sentences one to another, with their several Relations and Dependencies, to make a coherent Discourse' (III. vii. 1: 471).

[23] Here I echo Peter Geach's argument against Hobbes: 'Hobbes . . . held that the copula was superfluous; but we might very well object that it is necessary, because a pair of names is not a proposition but a list . . .' (1980: 60). (I do not think Geach's point in fact tells against Hobbes, since I think Hobbes has a more sophisticated view.)

[24] For more on this point, and the distinction between propositional unity and attitude, see Ott (2002).

this role.[25] And Hume himself treats both linguistic and non-linguistic cases together, speaking of signs 'in the countenance and conversation' of a subject.

So far Locke and Hume are very much in agreement: the claim that words are signs of ideas is to be read as the claim that words are reliable indicators speakers and writers give to their audience to allow them to infer from those signs to their ideas. But this is where the agreement ends.

24.4 JUDGMENT AND BELIEF

By far the most important difference between Locke and Hume lies in their treatments of what Locke called 'particles.'[26] It is striking that Hume never says that words can serve as signs of mental *acts* as well as objects. This is no accident, since he repudiates Locke's theory of judgment,[27] which itself was taken over from traditional Aristotelian sources. Locke, as we have seen, makes a sharp distinction between words that signify ideas and particles that indicate acts of the mind, which in turn allow us to combine or separate our ideas and so to form propositions. For Hume, by contrast, this is a serious mistake.

Hume begins his attack on the traditional theory by examining the sentence 'God is.' The traditional view is forced to read this existential claim as involving a judgment, and thus as involving at least two ideas, joined by the mental act corresponding to the copula. But Hume does not think that we have an idea of existence over and above the idea of the object said to exist (1. 3. 7. 2). This echo of Suárez's dictum that existence is not a predicate leads Hume to question the entire traditional way of understanding judgments. The 'vulgar' distinction of mental acts into conception, judgment, and reasoning is simply false; the latter two 'resolve themselves into the first, and are nothing but particular ways of conceiving our objects' (1. 3. 7. 5).[28]

The moderns sometimes use 'idea' when they clearly mean 'proposition' or 'judgment.'[29] Hume is at least not guilty of this kind of carelessness, since for

---

[25] A further argument could be made by looking, not at 'sign,' but at 'annex' and its cognates. In the case of Locke, for example, it is clear that he speaks indifferently of sensations signifying qualities, or qualities being annexed to sensations. See II. viii. 13: 136–7. Thus, the role of 'annex' at the level of ideas would provide a corollary to the role of 'signify' at the same level, buttressing the case for taking signification to be indication.   [26] See Ott (2006*b*).

[27] I am using 'judgment' in the technical sense and thus as Hume, but not Locke, uses it.

[28] See Ott (2004*b*: 146–7) and Owen (forthcoming).

[29] A word on 'propositions' in this context. I take this to refer to the sort of thing that can follow a 'that' clause. I mean to be ontologically neutral with regard to their ultimate status. And although there is a variety of uses of the term in the modern period, I think it's fairly clear that a very common, if not dominant, use corresponds to my own. Consider Hobbes's definition: 'A *Proposition* is a speech consisting of two names copulated, by which he that speaketh signifies he conceives the latter name to be the name of the same thing whereof the former is the name; or (*which is all one*) that the former name is comprehended by the latter' (*De Corpore*, I. iii. 2: 30; see *Leviathan*, IV. 46, and *Human Nature*, I. v). For his part, Locke claims that the mind, 'either

him, a proposition formed by the mind *just is* an idea, imbued with a degree of liveliness. Although proceeding from an important insight about existential predications, Hume's view is widely regarded as disastrous.[30] For the distinction between simple perception and judgment was not a bit of scholastic detritus; it is well motivated, at least within the broadly mentalistic framework common to Hume and his predecessors. How can an idea, like that of God, have a truth value?

One might reply on Hume's behalf by noting that, if ideas are the primary bearers of truth values, they can in fact be said to be true or false to the degree that they correspond with what they represent. So a lively idea of God is 'true' just in case there is a God. After all, on Hume's view, this capacity for bearing truth values is a key difference between ideas and passions: 'A passion is an original existence, or, if you will, modification of existence, and contains not any representative quality, which renders it a copy of any other existence or modification' (2. 3. 3. 5). By contrast, an idea, as an ectype, can at least correspond or fail to correspond with its archetype. True enough; but this is no answer to our objection. There are in fact two levels to the objection. We have been focusing on the problem of propositional unity and content: in virtue of what does an idea come to represent a state of affairs? But even if we waive the difficulty of propositional structure, it remains the case that ideas at least must be referred to, or taken to be representative of, states of affairs, before they can have a truth value. In other words, truth and falsity only come on the scene when we assert (or deny) a proposition, e.g., *that* a given idea corresponds to an object; Hume thinks these attitudes can only consist in differences in the vivacity of the idea in question. But if believing that God exists is simply having a particularly 'lively or vivid' idea of God, it becomes obscure how a theist and an atheist could be said to disagree. It is hard to avoid the conclusion that Hume, in this respect, represents a significant step backward in the philosophy of language.

However that may be, Hume's theory of judgment is clearly at odds with that of Locke. Although they agree that meaning requires that each word be annexed to an idea, Locke relaxes this stricture when he comes to discuss particles (III. vii) so he can draw a distinction between ideas and propositions. Hume makes no such concession. Moreover, Locke is sensitive to the difference between joining ideas in a proposition and adopting an attitude to that proposition. Some

---

by perceiving or supposing the Agreement or Disagreement of any of its *Ideas*, does tacitly within it self put them into a kind of Proposition affirmative or negative' (*Essay*, IV. v. 6: 576). This is how mental propositions are generated; verbal propositions inherit their meaning from these. And of course Hume himself often uses 'proposition' in just this sense. For example, at *E* 7. 4, Hume says he has 'endeavoured to explain and prove this proposition' (namely, the copy principle). And at *E* 4. 1, Hume writes, '*That the square of the hypotenuse is equal to the square of the two sides,* is a proposition, which expresses a relation between these figures.' Such examples could doubtless be multiplied.

[30] Owen (forthcoming) provides the most sympathetic treatment I have been able to find.

syncategoremata provide propositional unity, while others indicate the attitudes taken toward the proposition so generated (III. vii). Hume has no mechanism for marking this difference.

Having taken account of these divergences from Locke's view, we can now complicate our picture of Hume's position to accommodate general terms. There is no intrinsic difference between these kinds of words or the ideas they signify and particular words and ideas, of course; rather, the difference lies in the roles they play. '[G]eneral ideas are nothing but particular ones, annex'd to a certain term, which gives them a more extensive signification, and makes them recall upon occasion other individuals, which are similar to them' (1. 1. 7. 1). A general idea is a particular idea that has the power to call to mind any one of a given class of other particular ideas, where that class is determined by resemblance in a common respect. This two-tier signification in the case of general words means we need a more complex model for Hume's overall view. If we focus on words themselves, it is clear that they both (*a*) provide a symptom or reliable indication of an idea in the speaker's mind and (*b*) cause us to form another idea (besides that of the word itself). If we look at a general word, we find that, while it plays roles (*a*) and (*b*), the idea it causes us to form itself plays role (*b*), in virtue of belonging to a resemblance class. It is in this sense that a general idea can be said to 'signify' other ideas.

24.5 SEMIOTIC EMPIRICISM

What, then, is Hume's final position? If we need a tag for it, I suggest 'semiotic empiricism':

> (SE) Words are reliable indicators of the speaker's ideas and revive ideas in us; if the speaker does not have an idea, they are nonsense.

Although Hume clearly endorses SE with regard to nouns, the situation is more complicated when we turn to the copula. Recall that 'God is' does no more than indicate that a speaker has a lively idea of God; 'is' makes no independent contribution. At the same time, 'is' does have a function in this case: to indicate the relative liveliness of the speaker's idea. Nevertheless, it neither signifies an act of the mind nor joins the idea of God with an idea of existence.

Note that, even in the case of nouns, SE does not require that one have a particular idea in mind. Someone uttering 'dog' and thinking of a duck-billed platypus is not speaking the same language as us; it would seem strange to claim that the speaker is guilty of nonsense. Nonsense arises when a speaker, as it were, pretends to have an idea she does not, not when she departs from typical usage.

For Hume, convicting a philosopher of committing nonsense is a two-step process: one first shows that a given view could not make sense to us unless we had an idea of a given kind, and then one uses the copy principle to show that we could not have such an idea.[31] Arguably, Hume holds that full-blooded cases of nonsense are quite rare. For example, when someone like Locke speaks of extra-mental objects, Hume suggests that he typically is thinking of objects simply as sets of perceptions distinct from his own and not as anything 'specifically different' from ideas, because no one can think of anything specifically different, and, given a modicum of charity, Locke must have been thinking of *something*. A much firmer case is provided by Hume's treatment of such terms as force, necessity, power, and so on, as applied to objects:

in all these expressions, *so apply'd*, we have really no distinct meaning, and make use only of common words, without any clear and determinate ideas. But as 'tis more probable, that these expressions do here lose their true meaning by being *wrongly apply'd*, than that they never have any meaning; 'twill be proper to bestow another consideration on this subject . . .   (1. 3. 14. 14).[32]

Even Hume's first formulation in this passage allows that we have unclear and indistinct ideas when we speak of bodies endowed with powers, rather than no ideas at all. But he clearly thinks the more usual case is one in which we have the 'true meaning' in mind—i.e., the one specified by Hume's two definitions of cause.

To sharpen the consequences of SE for Hume's empiricism, it is worth glancing back at Locke. As we have seen, Locke would insist SE be revised to incorporate particles:

(SE′) Words are reliable indicators of the speaker's ideas or mental acts, or a combination of both; if the speaker does not have these ideas or perform the mental acts, they are nonsense.

SE′ is less stringent than SE. A word like 'substance,' for Locke, can be meaningful even if it is attached not to a single idea but rather to a complex of mental acts and ideas. 'That which lies behind and supports observable qualities' is perfectly in order, from the point of view of SE′, since each word signifies either an idea or a mental act. Locke is consistently ambivalent, however, about the precise status of such terms. He frequently calls the 'ideas' they signify 'obscure' or 'relative.' At the same time, he sometimes suggests that there is no idea there at all. Thus, Locke's first entry under 'Substance' in his index reads: 'S. no Idea of it.'[33] Such relative ideas are mere placeholders, indications of gaps in our awareness of the world. From the point of view of SE′, of course, there is no special problem with such words, since they can be linked with the mental acts and objects that are signified by the relevant description.[34] The difference

---

[31] In this paragraph, I am indebted to an anonymous referee.

[32] Quoted in Stanford (2002: 343); he calls the dilemma Hume offers here 'the Choice': we can either confess to committing nonsense or acquiesce to Hume's analysis of causation.

[33] *Essay*, 745.

[34] This is not to say, of course, that Locke can supply the ideas of support and being that enter into the description of substance. See Ott (2004*b*: 108–13).

between these words and those that signify a single idea marks the limits of our experience and knowledge, not of our ability to speak meaningfully.

The tension in Locke's work plays itself out in Berkeley's. At least at one stage in his career, Berkeley wishes to adhere to SE, which is considerably more restrictive than its Lockean counterpart. A striking instance of this is afforded by Berkeley's attack on the substance view of the mind in his notebooks. Berkeley writes, 'Say you the Mind is not the Perceptions. but that thing wch perceives. I answer you are abus'd by the words that & thing these are vague empty words wthout a meaning' (*Philosophical Commentaries*, sect. 581). Here Berkeley challenges the propriety of the syncategorematic term 'that,' a crucial ingredient in the definite description. (Later, of course, he changed his view.[35]) At this point in his development, Berkeley accepts what he (mis)construes as Locke's linguistic thesis: all significant words must stand for ideas. He then notes that 'that' and 'thing' do not seem to be correlated with ideas and uses this to attack the definite description account of the significance of 'mental substance.'

Where does Hume fit in this spectrum? Three features of his view are crucial: he endorses SE, suitably understood; he conflates propositions with ideas; and he holds the copy principle. All of this conspires to make definite descriptions unattainable. Hume must say that lacking an impression of a Cartesian self, causation as the realist conceives it, or an external world, there is no sense to be made of existential propositions involving these alleged ideas, for a proposition just is a lively idea.[36] Since 'no reasoning can ever give us a new, original, simple idea' (*E* 7. 8 n. 12), we are left with the materials provided us by experience.

24.6 RELATIVE IDEAS

If this account is right, we are faced with a difficulty. For although Hume rejects Locke's mechanism for thinking beyond the bounds of experience—namely, definite descriptions and the acts of the mind that make them possible—he does allow for relative ideas, ideas, that is, of a bare something standing in a given relation to other perceptions. This seems to relax his semiotic empiricism so as to in effect allow in the definite descriptions Locke availed himself of. If this were the case, Hume should not deny the existence of the external world, or Causation, but merely refrain from affirming it. For we do have, as the theory of meaning requires, an idea, albeit a relative one, to be the object of belief.

---

[35] See, e.g., the discussion of notions in *PHK* I. 142 and *Alciphron*, Dialogue VII, in Berkeley (1949–58: v).

[36] Obviously I should not be understood here to mean that Hume takes *all* talk of the self or causal power to be nonsense. He offers his own distinctive accounts of what such language amounts to, as we shall see. He does, however, convict certain philosophical views using these terms of nonsensicality.

In three places in the *Treatise* (1. 2. 6. 9; 1. 4. 2. 2; 1. 4. 5. 19), Hume speaks of our having a 'relative idea' that seemingly allows us to think beyond the bounds set by our impressions.[37] The inspiration for this is no doubt Locke's own use of the phrase.[38] Why could Hume not say that we have a relative idea of Causation that really characterizes things in themselves?

We should begin by noting that Hume never uses the phrase 'relative idea' in the context of causation or necessity.[39] All three uses in the *Treatise* are confined to the question of the external world; Hume drops the notion altogether in *Enquiry*. What is more important, the notion of a relative idea cannot help us to think of Causation. Causation is a relation; the notion of a relative idea, by contrast, takes a relation of which we have already formed an idea and applies it to new objects. The hypothesis of the external world might perhaps use the idea of causation to think of something lying behind and giving rise to our perceptions.[40] But the nature of the case prevents such a strategy from even getting off the ground where causation itself is concerned. For there what is at

[37] Hume does implicitly discuss a relative idea of causation in the *Enquiry*; see Hume's footnote at *E* 1. 7. 29. There, Hume writes, 'According to these explanations and definitions, the idea of *power* is relative as much as that of *cause*; and both have reference to an effect, or some other event constantly conjoined with the former. When we consider the *unknown* circumstance of an object, by which the degree or quantity of its effects is fixed and determined, we call that its power...' This, one might think, allows us to form a relative idea of power or cause. See also *T* 1. 3. 14. 27, where Hume says that he is ready to allow that there are unknown qualities, and 'if we please to call these *power* or *efficacy*, 'twill be of little consequence to the world.' In both passages, the unknown qualities or circumstances Hume speaks of would not be powers or causes in the realist's sense, although they might be in Hume's own. That is, it is entirely possible that an ideal observer would notice constant conjunctions among qualities to which we do not have epistemic access. But this is not what the causal realist is after, which is precisely why Hume cheerfully grants this possibility in the *Treatise*. (Thanks to Dan Flage and Miriam McCormick for drawing my attention to these passages.)

[38] See *Essay*, ii. xxiii. 3: 296; ii. xxiii. 2: 295; ii. xxiii. 15: 305; ii. xii. 19: 175; and i. iv. 18: 95.

[39] As Blackburn (1993) notes.

[40] This won't work, of course, since the idea of causation that Hume thinks we *do* have is one that connects perceptions that have been conjoined in previous experience. The external world (construed as the realist does), however, is by hypothesis not an object of perception. Kail (2007: 60) argues that Hume is a realist about the external world, where the 'external world' is just something other than perceptions that nevertheless resembles those perceptions: 'Rather than trade on the supposition of "unknown somethings," the realist should instead simply point out that the supposed external objects *resemble* perceptions.' This seems to me contrary to the whole thrust of *T* 1. 4. 2. Explaining the origin of the doctrine of double existence, Hume writes, 'Philosophers deny our resembling perceptions to be identically the same, and uninterrupted; and yet have so great a propensity to believe them such, that they arbitrarily invent *a new set of perceptions*, to which they attribute these qualities. I say, a new set of perceptions: For we may well suppose in general, but 'tis impossible for us to conceive, objects to be in their nature any thing but exactly the same with perceptions' (*T* 1. 4. 2. 56; my emphasis). That is, if external objects were to resemble our perceptions, they would have to just *be* perceptions, just as Berkeley argued. Hume thus endorses Berkeley's claim that only an idea can be like another idea. (As Hume says, Berkeley's arguments, the most prominent of which concludes that only an idea can resemble another idea, 'admit of no answer and produce no conviction' (*E* 12. 15 n. 32). Though most commentators focus on the second conjunct, the first is most important.)

issue is not what fills a certain gap in a relation (i.e., the nature or kind of thing that stands in a relation to something else), but that relation itself.

So it is useless to appeal, as defenders of the New Hume often do, to relative ideas when trying to establish the intelligibility of Causation. Recall again Broughton's characterization of Causation as 'some feature of objects that underlies' the constant conjunctions we observe.[41] But this relation of underlying, if it is anything at all, just is the relation of causation.

To leave the matter here, however, would be to miss an opportunity to clarify Hume's semiotic empiricism, and to make a first start at understanding Hume's argument from nonsense. First, we should note Hume's own ambivalence about relative ideas. In 1. 2. 6. 9, he writes, 'The farthest we can go towards a conception of external objects, when suppos'd *specifically* different from our perceptions, is to form a relative idea of them, without pretending to comprehend the related objects.' Most of the time, that is, we take objects to be either our perceptions themselves or independently existing perceptions; Hume will go on to show both of these to be non-starters. But, as perceptions, they at least are the kind of things with which we are acquainted; we simply suppose that they stand in new relations. A relative idea, it seems, allows us instead to think about, however meagerly, objects 'specifically different,' i.e., different in kind, from perceptions.

When Hume takes up the topic at greater length (1. 4. 2), he refers back to the hypothesis of objects specifically different from perceptions only to say that he has 'already shown its absurdity' (1. 4. 2. 2). Returning to the same point in 1. 4. 5, he says that we can conceive such objects by means of 'a relation without a relative' (1. 4. 5. 19). Again, this is not supposed to provide comfort to the defender of the external world, so understood. In the very next passage, he announces that we 'never can conceive a specific difference betwixt an object and an impression' (1. 4. 5. 20).

Why does Hume think that relative ideas are no help to the defender of the external world? Surely part of the answer must be the influence of Berkeley. When considering a similarly denuded concept of the external world, Berkeley writes,

if what you mean by the word *matter* be only the unknown support of unknown qualities, it is no matter whether there is such a thing or no, since it in no way concerns us; and I

---

[41] Or consider P. J. E. Kail's 'Bare Thought' of necessity. Kail takes the Bare Thought to be the thought of something that does precisely what Hume thinks nothing can do: allow us to infer a priori the effects of any given cause. Kail thinks that 'Hume's negative arguments equip us with the Bare Thought necessary to meet the semantic threat to realism' (2007: 84). That is, by showing that no idea could play the role an idea of (mind-independent) necessary connection would, Hume's arguments provide us with precisely that idea, though only as a 'Bare Thought.' While Kail uses Hume's negative arguments to give the Bare Thought its bare content, he rejects what he calls the 'short argument' for the incoherence of the idea of (real, mind-independent) necessary connection. But the negative arguments (2007: 84) and Kail's short argument (2007: 88) are, I think, at bottom just the same. So I take the negative arguments to show that no idea in principle could do what the realist wants, not that we happen to fail to have it.

do not see the advantage there is in disputing about we know not *what*, and we know not *why*.   (*PHK* I. 77)

As Simon Blackburn puts it, paraphrasing Wittgenstein, 'nothing will do as well as something about which nothing can be said.'[42] Nevertheless, unlike Berkeley, Hume thinks the idea of non-perceptual external objects is coherent, though empty. We can think of relations, which seemingly means we can form ideas of things standing in them, without specifying their nature.

The explicit reply to the defender of the external world comes from the doctrine of causation (1. 3. 14. 36): 'we can never have any reason to believe that any object exists, of which we cannot form an idea.' Since causal reasoning is the only way we come to believe that anything not immediately perceived exists, we must rely upon it alone in the case of the external world. But causal reasoning is simply a transition from one perception to another. Something that was in principle unperceivable, then, could never be thought of under the relation of cause and effect.

Hume, then, seems to think that the hypothesis of the external world, understood as specifically different from our perceptions, is (just barely) intelligible, but unsupportable. At the same time, Hume does *not* think that this skeletal conception is the object of an irresistible belief; it is instead purely a philosopher's fiction. What we cannot help believing is either that our perceptions themselves are possessed of continued existence (the belief of the vulgar), or that there is yet another set of perceptions lying behind our own. Why, though, are external objects (i.e., objects different in kind from perceptions) not the subjects of this irresistible belief?

Here is where we must keep in mind Hume's picture of belief. When I say 'the external world exists,' I am indicating the liveliness of an idea, which approaches that of an impression (1. 3. 10. 3). But what idea?

First, we should note that for Hume, any relation must be a relation between perceptions. What happens, then, when we transfer a relation from its usual home, and put perceptions in general on one side and nothing at all, or the bare notion of something, on the other? Remember that Hume says (1. 4. 2. 2) that he has shown the absurdity of supposing that external objects exist. Why does he think he has done so? He refers back to 1. 2. 6, where he describes this supposition as involving the notion of a relative idea. So the very notion or content of any merely relative idea involves an absurdity. We now have to figure out what this absurdity consists in.

A relative idea takes a relation between perceptions and tries to use it in a new context, one in which it does not belong. This does not guarantee that we have an idea after all. A relation is always a complex idea, which includes the ideas of its relata (1. 1. 5; see below). Now, an idea with a missing constituent cannot in

---

[42] Blackburn (1993: 103); see Wittgenstein, *Philosophical Investigations*, I §304.

fact be an idea. This is why, I think, Hume does not suppose we are irresistibly drawn to believe in external objects as specifically different from perceptions. There is no proper object of belief there.

We can support this reading by looking at Hume's use of 'supposition.' In 1. 2. 6, where Hume thinks he has established the absurdity of external objects, he speaks of such objects as 'suppos'd *specifically* different from our perceptions' (1. 2. 6. 9). As Winkler has noted, a common eighteenth-century use of 'suppose' allows one to suppose absolutely anything, even something that is logically contradictory.[43] Hume's talk of absurdity, I think, is in line with this: those who speak of external objects can suppose that they are different in kind from perceptions. But this is logically incoherent.

We are now in a position to put our finger on this contradiction at the heart of external objects. What is the bare notion of something, which the relative idea requires? The only notions of 'things,' 'objects,' or 'beings,' we have is that of perceptions. This is why we can never succeed in believing that there are specifically different objects out there: whenever we attempt to solve for *x*, to fill in the gap in the relative idea, we have to supply our notion of perceptions. It is *this* belief, not the belief in external objects, that we find irresistible. This is why, although we can suppose that external objects exist, we cannot so much as conceive of such objects.

Thus, the notion of a relative idea, I think, is Hume's admittedly awkward way of flagging the difference between a supposition—a way in which words can be combined so as to *look* meaningful—and a meaningful string of words. The three mentions of relative ideas in the *Treatise* do not, then, weaken Hume's semiotic empiricism.

The illusion of meaning by 'the external world' a set of external objects disappears as soon as one takes the time to move beyond words to ideas. This is hardly an unusual state of affairs. In considering space, and in particular the notion of the vacuum (matterless extension), Hume considers the objection that these words must be significant since they have been the subject of disputes (1. 2. 5. 2). Answering this objection at the end of that section, Hume writes,

The frequent disputes concerning a vacuum . . . prove not the reality of the idea, upon which the dispute turns; there being nothing more common, than to see men deceive themselves in this particular, especially when by means of any close relation, there is another idea presented, which may be the occasion of their mistake. (1. 2. 5. 22)

It takes some thought, in other words, to see that the supposition of extension without matter is incoherent. And when there is another idea in the area—in the case of the external world, the idea of perceptions—people are easily misled into thinking that they have used a word significantly, even though it stands either for that other idea, or no idea at all. As Hume puts it, once terms, however

---

[43] Winkler (1991: 560).

insignificant, get used often enough, 'we fancy them to be on the same footing with the precedent [i.e., words that are significant], and to have a secret meaning, which we might discover by reflection' (1. 4. 3. 10).

24.7 THE ARGUMENT FROM NONSENSE

Given this picture of Hume's restrictive semiotic empiricism, we are now in a position to see how it fits with his larger negative strategy. As I read Hume, his chief argument against the attribution of Causal powers to bodies is their unthinkability. This strategy is indirect in so far as it concerns not, in the first instance, the justification of an idea (or object of belief), but rather its putative origins. Yet it fits perfectly with Hume's overarching project.

That project, I have argued, has a tripartite structure: it begins with the quite general goal of tracing out the connections among perceptions and the origins of belief. At the second level, Hume proceeds to show that reason (both demonstrative and from experience) and sense cannot be responsible for a given kind of transition. Finally, to satisfy his initial goal at the top level, he shows how the imagination is ultimately the only explanation for that transition. Whether we choose to put this by saying that Hume denies that there is such a thing as reason where cause and effect are concerned, or, as I prefer, that Hume re-envisions reason so that it turns out to be something quite different, namely, a 'wonderful and unintelligible instinct' (1. 3. 16. 9), is a purely verbal affair.

The argument from nonsense is the key element in the second moment of this architectonic: there is, in principle, no idea of anything that could play the role of Causation in the realist's script. What I have done so far, however, is merely to sketch the argument in its most general form; the specific arguments Hume gives—the no necessary connection and epistemic arguments—have not yet been developed. Before doing so, however, I wish to get a bit clearer on the form of the argument itself. In particular, I want to investigate, first, whether Hume can give a coherent account of the mistake people make when they speak of something without having an idea of it and, second, how the argument closes off the in principle possibility of Causation and makes it impossible for us to posit Causation by means of an inference to the best explanation.

The argument from nonsense aims to show that a given speaker either utters gibberish or means something quite other than what she thinks she means. P. K. Stanford has dubbed this argumentative strategy 'the choice.'[44] We have already seen Hume offer his opponents this choice in the case of the self and the external world; he does the same thing with substance: 'we have . . . no idea of substance, distinct from that of a collection of particular qualities, nor have we any other

---

[44] See Stanford (2002).

meaning when we either talk or reason concerning it' (1. 1. 6. 1). But such collections of qualities 'are commonly referred to an unknown *something*, in which they are suppos'd to inhere' (1. 1. 6. 2). This Lockean notion of substance is no notion at all, but a mere concatenation of words. So here again we find an implicit choice: one can either confess to meaning by 'substance' nothing but a collection of qualities, or one can stand convicted of supposing something incoherent, that is, using words without ideas.

It is important to contrast what we might call the philosophical nonsense this practice of supposition generates with plain old Jabberwocky-style nonsense. In the latter case, sounds or letters are combined more or less at random. Philosophical nonsense is distinguished by its use of words that are perfectly in order in other contexts. We can use words that individually are significant to produce a grammatically correct phrase or sentence without seeing that we have no ideas, or, as Hume thinks is more likely, quite other ideas, lying behind them. The obvious objection is that this is simply incoherent: to spot such nonsense, I would have to understand it well enough to see that we could have no such corresponding idea, which means that those words are not, after all, nonsensical. Hume can reply, however, that spotting philosophical nonsense requires only that one understand the constituent words in other contexts, not that one understand the whole string. Take, for example, the phrase 'the square root of Bismarck, North Dakota.' No idea can be formed of such a thing. But I know this only because I know the meanings of 'square root' and 'Bismarck.' This does not mean that 'the square root of Bismarck' signifies an idea, even though, individually, each word does. Indeed, this is precisely what explains the status of this phrase as philosophical, as opposed to ordinary, nonsense.

Let us turn, now, to the implications of this for the in principle possibility of Causation. Now, in the coming chapters, I shall show just how Hume, like Malebranche, thinks our minds have a tendency to 'spread' themselves on to the world. In the case of causal power, this amounts to our making a kind of category mistake. The following chapters will show in greater detail just why necessity cannot be a feature of mind-independent reality. To close the present chapter, I want only to draw its obvious conclusion: that the features of Hume's view we have already explored close off the possibility of a world structured by Causation.

A typical contemporary response to Hume is to posit Causation as the object of an inference to the best explanation. This move is characteristic of the Dretske–Tooley–Armstrong view.[45] Armstrong, taking induction to be rational,[46] argues that it can only be so if we posit laws of nature that

---

[45] See esp. Armstrong (1983) and Dretske (1977).

[46] Armstrong holds that induction is partly constitutive of rationality, just as Peter Strawson does. Armstrong's view is distinguished from Strawson's, however, in that he offers an explanation for why this is so, whereas Strawson takes it to be an analytic truth. On Armstrong's view, induction

govern individual causal connections. These laws are the best explanation of the regularities we observe. In any individual case, the F-ness of a given particular entails its G-ness, because F and G, themselves universals, stand in relation N, necessitation. This entailment is not, for Armstrong, a logically necessary one; there are worlds where F-ness and G-ness are not so related. But in the worlds in which this relation holds, an instance of F-ness must be followed by an instance of G-ness.

Hume, of course, does not explicitly respond to this sort of picture. But we can see immediately that it is a non-starter, once one accepts the features of his view we have been exploring. An inference to the best explanation, where one has no idea of the explanans, is no inference at all. For lacking such an idea, all we can do is combine words in a way that can fool people into thinking we are saying something with content. Thus, Hume writes,

> If we have really an idea of power, we may attribute power to an unknown quality: But as 'tis impossible, that that idea can be deriv'd from such a quality, and as there is nothing in known qualities, which can produce it; it follows that we deceive ourselves, when we imagine we are possest of any idea of this kind, after the manner we commonly understand it. All ideas are deriv'd from, and represent impressions. We never have any impression, that contains any power or efficacy. We never therefore have any idea of power. (1. 3. 14. 11)

A parallel argument, of course, applies to Armstrong's view. What, precisely, is the N-relation that governs causal processes? Lacking an impression of it, we cannot form an idea; and since it is itself a relation, relative ideas, even if they *could* help in other cases, cannot help here.

I have been arguing, then, that a prominent contemporary response to Hume will not work *if one accepts* Hume's picture of intentionality and meaning. This is a fairly weak claim in one sense, since presumably few hold that view in all its details. But David Lewis, for example, makes similar points, and heaps Humean contumely on Armstrong's head.[47] Indeed, Armstrong's N-relation becomes all the more mysterious once one realizes its contingent nature. In virtue of what are ((necessarily) F-ness implies G-ness) worlds different from worlds where the F-ness–G-ness connection is a mere regularity?[48]

All I have done so far is to lay out the beginnings of the argument from nonsense. To make good on its claim that we have no means of representing Causation, Hume must go further. He cannot, and happily does not, simply appeal to his copy principle and his restrictive conception of linguistic meaning to show this. This is partly because the supposition (in the technical sense explained above) of Causation is the result, not simply of failing to pay attention to the

---

is rational because it derives from an inference to the best explanation, which leads us to posit laws of nature. And inference to the best explanation is, by definition, rational. See Armstrong (1983: 52–3).   [47] See Lewis (1983).

[48] Armstrong, of course, is willing to bite the bullet and admit that this is a brute fact.

contents of experience, but of a tangled web of other considerations. Thus, Hume deploys a variety of other arguments—including the 'no necessary connection' argument we have examined in Malebranche—to show that experience cannot provide us with the requisite impression to ground a realist conception of cause and effect.

The argument from nonsense, then, is best seen as a preliminary constraint on any account of causation. In the coming chapters, I will look to another constraint, which I call 'the practicality requirement.' Satisfying the first goal Hume has set himself means that certain stories abut causation, namely, those that swing entirely free of human (and indeed non-human) cognitive practices, are ruled out. That is, the notion of cause and effect has its home in our everyday reasonings, and those of non-human animals; any account that fails to respect this is not just entirely useless in the psychological project but for that very reason not an account *of* cause and effect.

Thus, it is only after these two main constraints—the limitations on representation imposed by the argument from nonsense and the practicality requirement—are on the table that we will be able to delve deeper into Hume's arguments against Causation. So in Chapters 25 and 26, I turn to Hume's positive accounts of necessity and relation. Here, the connection between the middle and bottom layers of Hume's strategy will become clear: his pictures of necessity and relation are meant as replacements for what he thinks are incoherent fictions.

25

Necessity

25.1 FINDING HUME'S TARGET

Like Malebranche, Hume takes Causation to involve *logical* necessitation; he deploys Malebranche's 'no necessary connection' argument (NNC). But the context of his larger project, I shall argue, gives his use of that argument a very different cast.

Hume writes,

Now nothing is more evident, than that the human mind cannot form such an idea of two objects, as to conceive any connexion betwixt them, or comprehend distinctly that power or efficacy by which they are united. Such a connexion wou'd amount to a demonstration, and wou'd imply the absolute impossibility for the one object not to follow, or to be conceived not to follow upon the other: which kind of connexion has already been rejected in all cases. (*T* 1. 3. 14. 13; see *E* 7. 7, examined below)

The first thing to notice here is just how different this statement of the argument is from Malebranche's. Malebranche set out to prove directly that no two finite objects or events, whether mental or physical, could be causally connected. Hume's conclusion is different: he thinks we cannot form an idea of such a connection. This is a product of Hume's quite different goals. Hume wants to know how minds function; settling this will have consequences for metaphysics, rather than the other way round. Thus, NNC is, for Hume, subsumed under the argument from nonsense: it provides us with further reason, beyond merely the stripped-down empiricism embodied in the copy principle and Hume's linguistic thesis, to doubt that we are equipped with an idea of Causation.

As with Malebranche, however, we must find the right target. I have established above that core features of Aristotelianism persist throughout the early modern period, particularly, of course, the understanding of causation as logical necessitation. Before proceeding, it is worth, I think, pausing over two more examples of this tendency.

Card-carrying Aristotelians were hardly swept off the face of the earth by the attacks of Descartes, Galileo, et al. Even in Britain at the very end of the seventeenth century, for example, we find John Sergeant, Locke's indefatigable critic, defending what is in many respects the scholastic view. What is novel is, *inter alia*, Sergeant's resistance to building the condition of God's concurrence into the causes themselves; he often sounds more like a conservationist. What is borrowed is much clearer: Sergeant, just as the defender of NNC anticipates, argues that for a cause not to produce its effect is a contradiction.[1] One is tempted to suppose that Sergeant is making an elementary mistake: of course, if one thinks of the cause and its effect under those descriptions, it is, trivially, necessary that the one produce the other. But Sergeant's point is precisely that of his Aristotelian predecessors, minus the concurrentism:

> For since all Truths are *taken from* the Nature of the Things, and from their Metaphysical verity, and consequently are in the Nature of the Thing fundamentally; and This is Contain'd and Exprest in the *whole* by Identical Propositions, and *in all its parts* by the Definitions; it follows that all Truths are Virtually contain'd in *Identical Propositions*, and, consequently, in the *Definitions*.[2]

Sergeant's 'identical propositions' are analytic in that they simply unpack what was contained in the subject of those definitions. Nevertheless, they reflect the real natures of things and not merely our intention to use words in a certain way. On Sergeant's view, the order of causes and the order of reasons differ only in that the former 'speaks the thing as it is in *Nature*,' and the latter, 'the same thing as 'tis in our understanding.'[3]

These real natures make it in principle possible for an ideal observer to deduce or infer the effects of its bearer. Although experience is required, on Sergeant's view, to sharpen one's understanding of that nature, this in principle possibility remains, just as Locke retains the possibility of predicting the key's ability to move the tumblers of the lock. As Hume puts it, 'were the power or energy of any cause discoverable by the mind, we could foresee the effect, even without experience; and might, at first, pronounce with certainty concerning it, by the mere dint of thought and reasoning' (*E* 7. 7).

But of course adherence to the general features of the Aristotelian view was not confined to those who identified themselves with the scholastics. When we turn to the work of Thomas Hobbes, we find him picking up the key point of the Aristotelians' thinking about causal necessity. Hobbes denies that we have anything approaching a science of nature in the Aristotelian sense. On his view, geometry is 'the only science that it hath pleased God hitherto to bestow upon mankind.'[4] In investigating nature, all we can hope for are 'probable' causes; that is, we can at best arrive at a hypothesis about causes in nature. This is partly because Hobbes agrees that causation is logical necessitation; settling on a true and complete cause would make it impossible to conceive of any state of affairs following it but its effect. Hobbes writes,

---

[1] See Sergeant (1696: 276).   [2] Sergeant (1696: 267).
[3] Ibid.   [4] *Leviathan*, I. iv, in Hobbes (1839–45: iii. 23).

a cause is the sum or aggregate of all such accidents, both in the agent and the patient, as concur to the producing of the effect propounded; all which existing together, it *cannot be understood* but that the effect existeth with them; or that it can possibly exist if any one of them be absent.[5]

The accidents or properties of the objects involved thus logically necessitate the occurrence of the effect. As Frithiof Brandt[6] and others have pointed out, Hobbes's definition of a necessary proposition is strikingly similar:

A *necessary* proposition is when nothing can at any time be conceived or feigned, whereof the subject is the name, but the predicate also is the name of the same thing; as *man is a living creature* is a necessary proposition, because at what time soever we suppose that the name *man* agrees with any thing, at that time the name *living creature* also agrees with the same.[7]

Just as we cannot conceive that a necessary proposition is false, we cannot conceive that a given cause does not produce its effect. Hobbes suggests that we can do our best to discern the aggregate of relevant accidents by a combination of both physical and thought experiments. We must observe repeated instances of an event in order to discover as many of the accidents present as possible and then do our best to determine *which* accidents of the alleged agent and patient are such that if they are present, we cannot imagine the effect not being produced:

we must examine every single accident that accompanies or precedes the effect, as far forth as it seems to conduce in any manner to the production of the same, and see whether the propounded effect may be conceived to exist, without the existence of any of those accidents; and by this means separate such accidents, as do not concur, from such as concur to produce the said effect; which being done, we are to put together the concurring accidents, and consider whether we can possibly conceive, that when all these are present, the effect propounded will not follow; and if it be evident that the effect will follow, then that aggregate of accidents is the entire cause, otherwise not; but we must still search out and put together the accidents.[8]

Thus, conceivability is used by Hobbes as a guide to causality, even though he does not take it to provide any *guarantee* that we have hit on the true cause. We begin with an empirical inquiry to help us refine our understanding of which accidents or properties of things are involved in producing the effect. We then test these hypotheses conceptually: can we conceive of a state of affairs in which all of these accidents are present and yet the effect is not produced? Given our ignorance, however, 'In natural causes all you are to expect, is but probability.'[9]

[5] *De Corpore*, I. vi. 10, in Hobbes (1839–45: i. 77; my emphasis).   [6] Brandt (1927: 274).
[7] *De Corpore*, I. iii. 10, in Hobbes (1839–45: i. 37–8).
[8] *De Corpore*, I. vi. 10, in Hobbes (1839–45: i. 77).
[9] 'Seven Philosophical Problems,' in Hobbes (1839–45: vii. 11). Other texts of Hobbes's, however, might be taken to support the claim that on his view divine omnipotence is a barrier

Charles McCracken has argued that Hume's treatment of causation as logical necessity is best explained by the influence of Malebranche.[10] McCracken has indeed shown that Malebranche was a potent force in the formation of the *Treatise*. But to leave it at that is to leave unexplained Hume's method of proceeding. It is not as if Hume happened upon some arguments of Malebranche's and simply took them over. What Malebranche considers the 'pagan' philosophy of the ancients, one that sees logically necessary connections in the world, persists throughout the modern period. Given its sometimes obscured or submerged presence at the heart of such disparate views as those of Régis, Sergeant, Hobbes, and Locke, it is no surprise that Hume takes it as his target.

25.2 AGAINST THE COGNITIVE AND GEOMETRICAL MODELS

I have been arguing that the scholastic view of causation is transformed into the cognitive and geometrical models. What both share with their common ancestor is the insistence on the logical necessity of causal ties. We can now ask just how effective Hume's arguments are against this triumvirate of positions.

We have already seen that Hume deploys Malebranche's own 'no necessary connection' argument. Its linchpins are his copy and separability principles: all ideas are copies of impressions and have no more content than do those impressions; and all distinct ideas are capable of being separated by the mind. Hume can then pose a dilemma: either the ideas of a given cause and its effect are distinct or they are not. If not, they are ideas of one and the same thing, and nothing can be its own cause. If they are distinct, they are separable, and hence, not necessarily connected. The question then becomes whether this is fair to Hume's opponents, and precisely how it is meant to work. We should begin with the Aristotelian view before moving on to the cognitive and geometrical models.

On the Aristotelian picture, the containment of a thing's powers in its essence means that its characteristic effects flow, as a matter of logical necessity, from that

---

to sublunary causation. Hobbes does indeed remark that 'There is no effect in nature which the Author of nature cannot bring to pass by more ways than one' (*Decameron Physiologicum*, in Hobbes 1839–45: vii. 88). That is, even after our observations and conceptual experiments, we might not have hit on the actual cause, simply because, although (necessarily) the complete cause brings about its effect, any given effect might follow from any number of distinct causes. This is not a result of God's omnipotence but a consequence of the general principle that 'same cause, same effect' does not entail 'same effect, same cause.' But Hobbes does not deny the existence of necessary causal connections that, if known, would render certain counterfactual states of affairs inconceivable.

[10] See McCracken (1983: 262): 'It is not in Locke, Berkeley, or even Descartes, that one finds emphasis placed on the indispensable role *necessary connection* plays in our idea of causation; it is Malebranche alone who lays great stress on the doctrine that causality consists essentially in a necessary connection of things.'

essence. Now, cause and effect are not identical, on this view: fire is distinct from the burning paper. That fire burns paper, however, is an analytic a posteriori truth.[11] If one envisions a scenario in which fire fails to burn paper under standard conditions, the Aristotelian might reply, one simply has not thought hard enough about the matter.

Now, this move will be of little help in performing our inductions. To be told that the colorless liquid I just ingested that made my skin burst into flame must not, after all, have been *water* will be cold comfort. Such a move threatens to make all inductions vacuously confirmable.

The temptation to make this kind of reply on Hume's behalf is itself of Humean origin. It is the stripped-down Humean picture of intentionality that comes to make this reply seem irresistible. For the Aristotelian believes that we do indeed have an ability to detect and reidentify instances of natural kinds. A properly functioning intellect will pick up on the real natural kinds out there in the world, and not merely in virtue of descriptions that capture only the superficial characteristics of this kind. Of course, it is precisely this ability that Hume questions, or better, whose possibility he simply cannot see. Such an ability is grounded in the Aristotelian's story about intentionality, and the existence in the intellect of an intelligible species abstracted from phantasms.[12]

The picture of intentionality that underwrote the Aristotelian view is obviously not Hume's. Humean impressions are a far cry from intelligible species. In fact, the separability principle is best understood as the direct outcome of the rejection of Aristotelian forms. For an Aristotelian form, once in the mind, would be precisely the sort of thing that *could* make it impossible for one to separate two distinct ideas, such as fire and burning paper.

This rejection is not unmotivated, but it isn't argued for, either. Given an atomistic reading of experience that confines it to our own sensations, the notion of a posteriori necessity becomes incoherent. Without the undergirding of an Aristotelian conception of intentionality, it becomes obscure how the forms of objects could become objects of thought. Consider Hume's claim that, if *a* and *b* are 'distinct,' there is no necessary connection between them, since the mind can conceive *a* and *b* separately. But unless we can give an independent sense to 'distinct,' the claim becomes trivial: distinct existences are separately conceivable by definition.[13] I take distinctness to be an irreducibly phenomenal concept, so that no two objects can be necessarily connected. What, then, makes them two and not one? Simply the different regions of qualitative space they occupy, and

---

[11] For the sense of 'analytic' at issue here, see sect. 3.1 above.

[12] I do not wish to leave the reader with the impression that I am arguing for an Aristotelian ontology or philosophy of mind. Intelligible species are no more plausible today than they were in the 17th century. A more promising line to take would be to appeal to the at once causal and epistemic individuation of properties envisioned by, e.g., Sidney Shoemaker (1980).

[13] An alternative way to put the same point is that Hume cannot produce a non-circular definition of 'distinct'; see Stroud (1977: 47) and Kail (2007: 93).

the phenomenal qualities with which they suffuse that space. The case is simplest when we focus on a single sense modality, such as vision. Two (visual) objects are different just in case they occupy different regions of the visual field; this should not be surprising since, for Hume, objects are perceptions.[14]

But it is not simply Hume's unquestioned assumption that all experience is of mental contents and not the world that drives the rejection of forms. A form points to its characteristic effects. And nothing can do this. Hume explicitly denies that an idea of a cause could point out or allow one to infer its supposed effect (*E* 7. 7). On its face, this is simply an assertion with little to back it up. Sergeant, for example, could reply that our typical failure to infer an effect from a single cause owes more to the unrefined notion a single experience gives us than to the paucity of the form that notion mirrors. To understand why Hume thinks it is in principle impossible to infer an effect on the basis solely of the idea of the cause, we need to keep in mind his contrasting and sparse picture of intentionality.

We can get at this point from another direction if we turn to the geometrical model of causation developed above. There, the idea was not that scholastic forms necessitate their effects, but that the geometrical properties of bodies make certain results inevitable. I have already argued that this development was the result of preserving the scholastics' emphasis on logical necessity in the context of ontological mechanism. This is precisely the mistake Hume diagnoses. 'Solidity, extension, motion; these qualities are all compleat in themselves, and never point out any other event which may result from them' (*E* 7. 8). This denial of physical intentionality is, I believe, another side of NNC: the kind of pointing Hume denies here just is the logically necessary connection that would allow us to demonstrate a causal sequence.

Nor does Hume share Régis's and Locke's faith that mechanical explanations stand on their own two feet. The *esse-ad* of scholastic powers is no less mythical when adapted to the constraints of ontological mechanism. These prior figures think that explanation has to come to an end; in the case of bodily events, it stops with their geometrical properties. What more, Régis wonders, could one ask for? But Hume simply cannot see how one property could 'point to' another.

However amenable it might be to contemporary anti-realist prejudices, Hume's use of NNC is, like Malebranche's, in danger of circularity. It is persuasive only when one accepts certain background assumptions, which are no less controversial than the conclusion itself. Like many other thought experiments, it seems to show us what we would have to be committed to in order to hold a given view, and not what we *should* be committed to.

I think Hume's NNC is best seen as an episode in his argument from nonsense, that is, an argument directed at showing that we have no idea of Causal power. If we had such an idea, certain states of affairs would not be conceivable that he

---

[14] See e.g. Kail (2007: 93, 134–5).

thinks plainly *are* conceivable. As in Malebranche's case, everything here turns on the content of what one is conceiving, and how it is described. Hume is free to describe it in terms of his crude and reductive picture of intentionality; is not the Aristotelian, or Locke for that matter, free to describe it in other terms, and hence to treat the conceivability of those states of affairs Hume insists on as merely apparent?

Recall that, for Locke, the truths that underwrite physics are synthetic and a priori. He would agree that there is no identity between a cause and its effect; rather, the connection is one built on the intrinsic intelligibility of the truths of geometry, which cannot be reduced to a containment between forms, as the Aristotelian wished to do. To challenge this, Hume appeals not only to NNC, but to his own account of necessity. The distinction between the positive and negative aspects of Hume's treatment of causality, which I have been using as a heuristic, is to this extent misleading. For Hume's replacement for the geometrical model is at once a diagnosis and a key element of his argument against it. A full account of this replacement will be developed below; thus, a final assessment of Hume's overall argument must wait.

For now, we can turn to Hume's attack on the cognitive model. Just as NNC is subsumed under the argument from nonsense, so is Hume's challenge to the occasionalism of Malebranche. If we cannot get an idea of causal power from our own experience, deriving it from God's activity is a non-starter (*E* 7. 25). (In fact, it is just the same as attributing Causation to unknown qualities, something that should make those friendly to the New Hume reading think twice. If lacking an idea of causal power prohibits us from attributing this power to God, it should equally prohibit us from attributing it to bodies.) Hume's ideas and impressions supplant not just the Aristotelian framework but Malebranche's own doctrine of the vision in God. To say that God's will is necessarily connected with its effect is, in the context of Hume's picture of cognition, mere stipulation, or 'supposition.' One would have to have some independent experiential grip on God and his activities to underwrite such an attribution; at the very least, one would have to have a grasp of how our own minds initiate actions.

This, too, Hume denies, and he uses one of Malebranche's own arguments, the epistemic argument to support it. Unlike Malebranche, Hume subordinates the epistemic argument to the argument from nonsense, and even then offers it only in the *Enquiry*.

This feature—Hume's emphasis on experience as unable to provide, even by introspection on volitions, an idea of necessary connection—makes his epistemic argument really quite different from Malebranche's. Malebranche's point in the epistemic argument was not, *pace* McCracken,[15] that we get no idea of causation from inner experience, though of course he would agree; instead, Malebranche's argument was aimed at showing that minds cannot be causes. Thus, Hume does

---

[15] See McCracken (1983: 261).

not face the challenge Malebranche did. Above, we saw that it took some work to make sense of Malebranche's argument, since common sense suggests that to bring about a given effect, where this can be done only by means of intervening events, it is enough to initiate those events and then walk away. Hume, by contrast, does not face this objection, because his point is quite different. As Hume states the argument:

It must be allowed, that, when we know a power, we know that very circumstance in the cause, by which it is enabled to produce the effect: For these are supposed to be synonimous. We must, therefore, know both the cause and the effect, and the relation between them. But do we pretend to be acquainted with the human soul and the nature of an idea, or the aptitude of the one to produce the other?   (*E* 7. 17)

The problem Hume diagnoses with minds as causes, then, is not at all the same problem Malebranche does. In the latter case, the difficulty was to see how a chain volition could achieve its effect when the first element in the series was unknown. Here, the problem is that inner experience affords us no idea of Causation; if it did, we would have to know how minds act, for these two are really just one and the same.

Hume, then, does not adopt Malebranche's cognitive model of causation. He does not rule out finite minds as causes on the basis of the ineffectuality of chain volitions. His argument here is, once again, an element in the argument from nonsense.

25.3 THE NEIGHBORING FIELDS

Neither outer nor inner experience equips us with an idea of necessary connection. This statement itself is apt to cause confusion, for of course there is a sense in which inner experience *does* provide an impression of necessity, and hence an idea of it. In this chapter, I shall speak of 'Necessity' when I mean necessity construed as a real feature of the (inner or outer) world, and 'necessity' when I mean Hume's proxy, which I shall explore below.[16]

That necessity in at least one of these senses is essential to causation is something Hume does not doubt. Hume, for example, would agree with the common contemporary view that causal claims commit one to counterfactuals.[17] If I say that *x* causes *y*, *ceteris paribus* I commit myself to the claim that if *x* had

---

[16] By the claim that necessity is a 'real feature,' of minds or of the world, I mean the claim that there is no possible world in which the events so related occur independently. Thus, if objects and events *x* and *y* are related by necessity so construed, there is no possible world in which *x* exists and *y* does not.

[17] See, e.g., Armstrong (1983) and Dretske (1977). Of course, these figures don't stop there: the conversational commitments one undertakes in making causal claims are merely part of the prima facie evidence they produce for the desirability of a realist account of causation (namely, as something more than constant conjunction).

not happened, *y* would not have, either. This, of course, needs to be sharpened up considerably to stand any scrutiny. But the basic insight seems to me, and to Hume, to be right: if I believe my hitting the window with a bat caused the glass to break, I also believe it would not have broken (then, and in those circumstances, construed in the nearest possible way to those of the actual world) had I not hit it. This is just the flip side of induction: I also believe that the next time I hit a (similarly structured) window with a (similarly constituted) bat, the glass will shatter.

The question is, what is the source of this commitment? Is it rationally justified? If so, it would require precisely the kind of idea of Necessity Hume has ruled out. But what, then, is its source? As Hume poses the question: does the inference from cause to effect, or vice versa, which takes place in both counterfactual and inductive contexts, depend on the idea of necessary connection? This is the assumption Hume thinks his forebears have been operating with. It has come to nothing.

It is at this stage of the *Treatise* that Hume announces he will 'beat about all the neighbouring fields' (1. 3. 2. 13). It will be another twelve sections before he solves his original problem. In the intervening ones, he develops the account of intentionality, meaning, and belief I have covered above. The neighboring fields, then, include the limits of human cognition, and the bounds of sense. What is vital for our purposes now, however, is that they also include Hume's thoughts on inductive inference. If all distinct ideas are separable, then nothing in them can account for our associating them; what does so must be found in the structure of our own minds. Thus, another possibility suggests itself: that the idea of necessity depends on the inductive inference. The inference is not itself justified; it is better described as a transition between ideas. But this transition, carried out often enough, can give rise to an impression and thence to an idea.

At first sight it is puzzling to see Hume switching gears in this way. He seems to have begun with a metaphysical question—what is the nature of causation?—and ended up with a psychological one—how do we infer new effects on the basis of experience? But this connection between the metaphysical and the psychological is precisely what we should expect, if I am right about the tripartite structure of Hume's strategy. Hume begins by asking after the origin of our idea of causation; since there is no single quality that characterizes everything we call a cause, Hume decides it must be a relation (1. 3. 2. 5).[18] The relations of contiguity and temporal priority are easily discoverable, and form, Hume thinks, part of our idea of cause and effect.[19] But the key element is missing. As

---

[18] Hume's argument has a weakness, since it does nothing to rule out the possibility that causation is indeed a single quality, though one that is multiply realizable.

[19] Some caution is needed on contiguity: see *T* 1. 4. 5. 12. Temporal, rather than spatial, contiguity seems necessary for Humean causation. And while the definitions of cause in the *Treatise* include both spatial and temporal contiguity, those in the *Enquiry* omit the reference to spatial contiguity. (I owe these points to Dan Flage.)

many philosophers have argued, thinking they were disagreeing with Hume, the common-sense, default conception of causation is not merely that of regularity, or constant conjunction. Such a conception leaves unexplained our tendency to affirm counterfactuals, and, as Hume points out, our tendency to infer similar effects from similar causes.

Even though in 1. 3. 2 Hume has not yet offered his suite of arguments against Necessity (this comes later, especially in 1. 3. 14), Hume states their conclusion:

When I consider their [objects'] *relations*, I can find none but those of contiguity and succession; which I have already regarded as imperfect and unsatisfactory . . . We must, therefore, proceed like those, who being in search of any thing, that lies conceal'd from them, and not finding it in the place they expected, beat about all the neighbouring fields, without any certain view or design, in hopes their good fortune will at last guide them to what they search for.   (1. 3. 2. 12)

In this way, Hume's top-level project leads to the middle level, where he argues that a given idea cannot have its source in reason or sensation. He then proceeds (1. 1. 5. 6) to investigate new ground, namely, the only remaining mental faculty: imagination. The second stage is indeed partly ontological in nature, since the real existence of something could, of course, help explain our having an idea of it. But when these fail, we need to turn to a different psychological mechanism to explain the mental phenomenon in question.

Hume never doubts that we infer effects from causes and vice versa. But if such inferences are not based on an idea of Necessity, perhaps the idea of necessity instead has its source in the inferences. Induction is important to the investigation of causation because causal reasoning is something we do: whatever the notion of causation is, it must have its source in actual cognitive practices.

25.4 THE PRACTICALITY REQUIREMENT

The practicality requirement simply formulates this constraint. No philosophical account of causation that swings free of actual cognitive practices can be right, not because these practices have a sacrosanct status as the arbiter of metaphysics, but because the concept of cause has its home in those practices.[20]

This requirement is at the heart of the new tack Hume takes in part 3 of the *Treatise* (1. 3. 2. 13) as he beats about the neighboring fields. One result of this

---

[20] It is not only the concept of causation that resists analysis unless we pay attention to its context in the psychological economy. Hume writes, 'We may also observe . . . in this instance of sounds and colours, that we can attribute a distinct continu'd existence to objects without ever consulting REASON, or weighing our opinions by any philosophical principles. And indeed, whatever convincing arguments may fancy they can produce to establish the belief of objects independent of the mind, 'tis obvious these arguments are known but to very few, and that 'tis not by them, that children, peasants, and the greatest part of mankind are induc'd to attribute objects to some impressions, and deny them to others' (*T* 1. 4. 2. 14).

reorientation is the observation that humans make causal inferences on the basis of repeated experience, not on the grounds of a single conjunction. This provides the clue: whatever it is that makes the difference between these cases will also be the missing element Hume needs, since even a single *a–b* sequence can of course exhibit temporal and spatial contiguity.[21] We can call this 'the Difference': whatever it is that distinguishes repeated observations of constant conjunctions, which produce an association of perceptions, from one-off observations, which do not.

This limitation is already fairly restrictive. Having reeled the notion of causation in, and characterized it as the idea crucial to induction, Hume has already ruled out any philosophical account of the notion that severs it from this background. But Hume goes much further and requires that any plausible account of causation be able to explain the kinds of reasonings in which non-human animals engage.

Hume thinks it is as obvious as anything can be that non-human animals also 'reason' causally (1. 3. 16). A bird sitting on an egg until it hatches and a chemist mixing up methamphetamine are not doing anything different in kind: 'As to the former actions [i.e., those of non-human animals], I assert they proceed from a reasoning, that is not itself different, nor founded on different principles, from that which appears in human nature' (1. 3. 16. 6). The kind of foresight exhibited by such animals can only rest, Hume thinks, on the same kind of inductive, and hence causal, inferences we humans make.

Thus, rather than continue to spin yet another metaphysical account of causation, Hume thinks we are better off pursuing the psychological project of isolating the element responsible for the Difference.

Why must this project be a psychological, as opposed to straightforwardly epistemological, one? As we have already seen, Hume has ruled out the possibility that anything in the perceptions themselves, or their relations, could be responsible for the Difference. The Difference must turn on a change in the mind, not in the mind's objects. It should be pretty clear that such a change will not in fact justify, in the typical sense, any causal inference. The Difference is a difference *not* in the things themselves but only in our reactions to them. And even this requires comment: it is not as if in repeated experience we discover some new feature *of the object* we had previously been ignorant of. Instead, repeated experience triggers a disposition in the mind. We are no more justified in believing that an unsupported object will fall to the ground than Helga is when she believes that my putting my shoes on means she will be taken for a walk. In neither case does the epistemic agent discover some intrinsic feature of the object or event that imbues the inference with justification. Having ruled out the possibility of justification, Hume turns to explanation instead.

[21] See *E* 7. 28.

26

Relations

We are now in a position to move from the second layer of Hume's project—the destructive one—to the final layer, which seeks to replace the vanquished accounts with an appropriately descriptive, psychological story. As I have said above, this replacement also functions as an argument against Causation; once one sees how the concept of cause actually works in its home environment, one will also see how silly it is to pretend that anything remotely like Causation is intelligible.

The task of this chapter, then, is to zero in on an analysis of causation that both meets the practicality requirement and does not violate Hume's restrictive picture of intentionality and meaning. To do this, we first have to look at Hume's treatment of relations and place it within its broadly Lockean context.

26.1 THE STATUS OF RELATIONS

With this chapter, we return to one of the main themes of the book: the ontological status of relations. We saw Locke and Boyle developing a view on which relations are not real items in the world; Locke, in particular, is quite clear that a relation is a mental act of comparison and so not of the right category to have a mind-independent existence. Nevertheless, both figures hold that some things in the world, and not merely our ideas of them, are fit subjects for comparison. I called this position 'conceptual foundationalism': while no relation is real, there is a real foundation in objects to justify their being compared in a given way. This is clearest with intrinsic relations like resemblance: that x and y have features F and F′, where these are taken as tropes (real universals having already been ruled out), their resembling each other is an objective fact, even though there is no element corresponding to 'resembles'; there are only the relata. Hume is barred from holding the view in this form. He thinks relations hold at most among perceptions; there is nothing in his account corresponding to Locke's semiotic epistemology that would allow him to say they also hold of mind-independent things, 'things' having been reduced to impressions. This is tied up with Hume's rejection of the external world; there is nothing really 'out there' for relations to relate.[1] But his positive account of the status of relations remains an open question, which I now propose to try to answer.

The best way to investigate Hume's own view is by posing a problem. On one hand, he explicitly says that no perceptions are necessarily connected. This will be familiar from the separability principle and Hume's use of NNC. But he goes further in the section on personal identity: 'All these [particular perceptions] are different, and distinguishable, and separable from each other, and may be separately consider'd, and may exist separately, and have no need of anything to support their existence' (*T* 1. 4. 6. 3).[2] Nothing, as it were, knits together the perceptions of the mind. The activity of the imagination is required if perceptions are to 'hang together' in coherent trains of thought and experience. But considered apart from this activity, each perception is 'entirely loose and separate.' When we compare perceptions, or make a transition from one to another, we generate (or use) a complex idea of a relation. Without this mental act, no perceptions are related.

At the same time, this seems to fly in the face of Hume's own way of speaking. Throughout the *Treatise*, Hume speaks of our 'discovering' some relations (see, e.g., 1. 3. 2. 2).[3] One cannot discover what one has created. From this sort of evidence, Alan Hausman concludes that Hume 'gives relations, however hesitatingly, objective ontological status.'[4] Better evidence comes from Hume's claim that objects are related by succession and contiguity, and that 'all this is independent of, and antecedent to the operations of the understanding' (1. 3. 14. 28). For his part, Wayne Waxman has relied on the separability principle and the section on personal identity to insist on just the opposite: perceptions cannot *really* be related, on pain of violating one of Hume's core principles. The thought seems to be that a real relation between *a* and *b* would make it the case that *a* and b are inseparable in thought; but if we really have *a* and *b*, and not just *a* and *a*, then this separability is always there.

The way out of this paradox will by now be clear. We can reconcile both sets of texts if we attribute to Hume a version of conceptual foundationalism. Consider a key text Hausman appeals to, in arguing for the mind-independent reality of Humean relations: 'It may perhaps be esteem'd an endless task to enumerate all those qualities, which make objects admit of comparison' (1. 1. 5. 2). Hausman is right that to speak of perceptions 'admitting' of comparison has a realist ring to it; and yet Waxman seems equally right to say that if perceptions were really related, they would be inseparable.

On the conceptual foundationalist view, modified to suit Hume's ontology, one can say that perceptions admit of comparison in so far as they are suitable

---

[1] For a different view, see Kail (2007).

[2] Also quoted in Waxman (1994). Waxman makes a good case for embracing this horn of the dilemma; see esp. (1994: 78 ff.).

[3] Other evidence for Hume's apparent realism about relations is assembled in Hausman (1967).

[4] Hausman (1967: 260).

objects for the mind to compare in a given way. This does not mean that those perceptions, however closely they might (for example) resemble each other, are inseparable, or that, without the activity of the imagination, one would somehow give rise to the other. Our problem then disappears, and Hume arrives at a consistent view. Moreover, the Lockean picture had become something like common sense by the time Hume was writing; it is enshrined, for example, in Ephraim Chambers's widely read *Cyclopaedia* of 1728.[5]

Thus, we do not need the desperate remedies that would be required to embrace either horn of our dilemma. This is not to say, however, that the nature of the 'fitness' of perceptions to be compared, or the justification of our doing so, is always to be found in the relata themselves. In some cases, this is so; as we have seen, four of the philosophical relations have their foundations in the perceptions themselves. In the others, however, their fitness is a matter not so much of the perceptions but of the manner in which they are introduced to the mind. This is true of the central case of causation, and the necessity that lies at its core.

What I have done so far, however, is merely a preliminary account. As we shall see, it will have to be much modified to accord with the rest of Hume's view. Ultimately, we will find that Hume has removed the justificatory aspect of conceptual foundationalism and replaced it with an account conducted purely in descriptive terms. This will then position us to understand his positive account of causation, and its two definitions.

26.2 TWO KINDS OF RELATIONS

The word *relation* is commonly us'd in two senses considerably different from each other. Either for that quality, by which two ideas are connected together in the imagination, and the one naturally introduces the other . . . or for that particular circumstance, in which, even upon the arbitrary union of two ideas in the fancy, we may think proper to compare them.   (1. 1. 5. 1)

'Relation' is ambiguous, though in either sense, a relation holds only between or among perceptions, not things. When we say that 'x is related to y,' we might mean that x and y are joined in the imagination by the 'gentle force' of attraction. In this sense, to say that x and y are related is to make a self-report: it is to indicate to others the connection one feels when contemplating the two ideas. (This will be vital below.) When Hume speaks of the 'quality' by which the two ideas are connected, he means the activity of the imagination as embodied in one of the relations discussed in the preceding section (1. 1. 4), namely, resemblance, contiguity, or cause and effect.[6] These are the 'natural' relations.

---

[5] See the entry under 'Relation,' in Chambers (1728: 988 ff.).
[6] Hume thus follows Locke and Boyle in sometimes calling a relation a 'quality.' See *T* 1. 1. 4. 1.

On the other hand, we sometimes use the word 'relation' to indicate, not this subjective psychological principle, but 'the circumstance,' or, perhaps better, the respect in which we compare perceptions, without any regard for a psychological connection by which the one introduces the other in the imagination. In this broader use, we simply ignore the 'quality' or associative principle that connects the ideas and think only of the ideas themselves in a certain way. It is in this sense that we may say that any two ideas whatsoever resemble each other, for example, even though the thought of one need not bring about the thought of the other. Relations, in this sense, are best thought of, not as arbitrary or voluntary, but as concerning the perceptions considered in the relation, as opposed to the psychological mechanism that connects them.

After all, as Hume says, the principles of attraction he discovers in 1. 1. 4 are themselves the source of our complex ideas of relations (1. 1. 4. 7). The philosophical relations discussed in 1. 1. 5, then, must ultimately be the result of the activity of the imagination, and, in particular, its disposition to compare ideas under the heading of the *natural* relations, which just are the principles by which ideas are connected. Thus, the natural relations are clearly prior to the philosophical: without the former, the latter would be impossible.[7] Unless the mind had a blind impulse to move from one idea to another when their objects were (immediately) contiguous, we would never think to compare two quite distant perceptions (see 1. 1. 5. 3).

It is tempting to think of philosophical relations as excluding, rather than simply remaining silent on, the associative principles instantiated by the three natural relations. But this would leave a serious gap in Hume's view. After all, philosophical relations require a mental act of comparison by which the ideas of them are generated. What would explain why this act takes place, if philosophical relations did not themselves depend on natural ones? As I see it, philosophical relations and natural relations are not two distinct kinds of things; rather, they are two different ways of talking about what is, in some cases,[8] one and the same thing. Every relation is a philosophical relation, simply because any association of ideas *a* and *b* can be considered by looking at *a* and *b* themselves and not paying attention to the fact of their association in the imagination. Not every relation, however, is a natural one, since we are free to compare any two ideas at all, even when we are not naturally or irresistibly led to think of one on the basis of the other.

It can be quite hard to see the difference between a 'quality' (natural relations) and a 'circumstance' (philosophical). It will be most stark in a case where the belief that a philosophical relation holds between x and y does not include, rather than simply ignores, the connecting principle or quality by which the imagination operates. As I have already indicated, this must be a matter of degree,

---

[7] As Waxman has persuasively argued; see Waxman (1994: 80 ff.).
[8] Namely, resemblance, contiguity, and cause and effect.

since without *some* operation of these natural connecting principles, however faint, no mind would ever in fact compare x and y. To say that there is no such connecting quality here is just to say that the mind is not always, or even typically, led to form the idea of y after thinking of x. Take the feature, then, of having a nose. This feature is so common that, upon seeing Louise's nose, I do not then immediately form the idea of Bob's; presented with too great a choice of which ennosed individuals on which to think, the imagination is prevented from seizing on any of them (1. 1. 5. 3). So to say that Louise and Bob resemble each other in that each has a nose is a paradigm case of a *merely* philosophical relation. Their ennosedness, we might say, is not a quality that connects any ideas in such a way that I am forcefully led from one to the other. At the same time, it is not a mystery that I can, if I wish, enumerate the ways in which Louise and Bob resemble each other, and begin my list with the fact that each has a nose. I have identified a circumstance in which they are related; I have not identified a quality that leads me (always, or typically, or with any great force) to think of one when I have thought of the other.

This comes out, I think, in the conversational commitments one undertakes. Suppose I say, 'Bob and Louise are very similar in appearance.' Now suppose I am challenged to defend my claim. I might say, 'well, they both have nine deep lines in the forehead.' This would explain why I picked out Bob and Louise, and not just any two human beings at all. This close similarity in a fairly distinctive respect explains why my thoughts began with Bob and ended with Louise, as it were. Suppose I say, instead, that 'they both have noses.' In one sense this is a perfectly fine answer: they do indeed have noses, and this makes them similar. But precisely because this circumstance is so widespread, I have given no particular reason for picking out these two people rather than any others. So even if my claim is quite true, it is a very odd conversational move. And it is odd precisely because we expect that most minds, most of the time, will be forming their ideas on the basis, not just of resemblance, but of the kind of close and particular resemblance that typically moves our minds to make the transition from one idea to another. Although this is a departure from Hume's language, it captures his basic method of distinguishing between the two kinds of relation. If I am right, we are well on our way, despite appearances, to understanding Hume's doctrine of causation.

We can see this if we reflect on the structure of part 3 of the *Treatise*. Now, all relations are philosophical; some are also natural. The philosophical are the wider class because they mention one fewer fact than do the natural: that the mind is led from one relatum to the other in a forceful and hence usually successful way. Investigating causation as a philosophical relation in 1. 3. 2, Hume throws up his hands and sets off for the neighboring fields. But he does so only when he comes to the question of necessity. Some analytical progress, in terms of the qualities of contiguity and temporal priority, has been made, but what baffles him here is the necessity that forms the core of our concept of causation. This is not at all puzzling, once we remember that 1. 3 starts off with a treatment of *philosophical* relations, and that it is *qua* philosophical relation that Hume is now considering the relation of cause and effect. Of course the idea of necessity goes missing if we consider causation in this way. For to treat it as a philosophical relation is precisely to ignore whatever connections might be set up in the mind by repeated observations and hence be responsible for the Difference. This will be important later, when we come to the vexed question of Hume's two definitions of cause.

26.3 THE NATURE OF NECESSITY

The Difference, then, is really a difference in our psychological states. On repeated experience of a conjunction, custom or habit allows us to associate the conjoined, so that the one perception 'determines' us to form the other. Thus, we have an impression of necessity: an impression of the mind's determination to make this transition. But what precisely is Hume's account of necessity?

We can begin by distinguishing two kinds of necessity: that which attends intuitively certain propositions, such as those of mathematics, and that which is involved in causation. In the former case, we simply look at the ideas in question, and, as Hume puts it, 'are determin'd by reason to make the transition' from one idea to the other (1. 3. 6. 4). In the latter, we are equally determined, but here by 'a certain association' of perceptions. Thus, it looks as if, at this relatively early stage of his investigation, we have two quite different mechanisms that allow or, better, force us to think of ideas as standing in certain relations: demonstrative reason, which detects foundations in the ideas themselves, and reason from experience, founded in the contingent, extrinsic 'association' of perceptions in experience.

But by the time we reach Hume's own account, the distance between these has evaporated. In a famous (or perhaps notorious) passage, Hume writes:

Thus as the necessity, which makes two times two equal to four, or three angles of a triangle equal to two right ones, lies only in the act of the understanding, by which we consider and compare these ideas; in like manner the necessity or power, which unites causes and effects, lies in the determination of the mind to pass from one to the other. (1. 3. 14. 23)

In either case, then, necessity is merely a characteristic or description of the mind's determination in its actions; it is not a real feature of the world, or a real connection between ideas. It seems odd, however, for Hume to lump these cases together. One can imagine an opponent being initially quite pleased to see causal connections raised to the same level of mathematical ones: what more could one ask for, after all? The pleasure disappears once one sees that Hume's point is not *pro* causal necessity but *contra* mathematical and geometrical necessity. That is, the necessity that characterizes mathematics is equally a matter of psychology.

What has become of Lockean conceptual foundationalism? Couldn't one admit that necessity, as a relation itself, is unreal, and yet insist that there is some basis for our formation of an idea or impression of it? This was surely Locke's view.

Precisely here—in bridging the gap between Locke's conception of relations as unreal but justified and Hume's—is where we find the payoff for having struggled through Hume's distinction between philosophical and natural relations. Necessity does not appear on either list. This is because it is what we might call a meta-relation: it describes the first-order relations that Hume catalogs. Relations like mathematical equality and causality can also be characterized at this meta-level. This will make matters more complex, since necessity, it would seem, figures in causal relations at the first level as well.

As a relation itself, necessity can be taken either as philosophical or as natural. The impression from which the idea is derived is the felt determination of the mind to move from one idea to another; what kind of ideas these are, and the respect in which they are being compared, determines the kind of necessity—causal, mathematical, or what have you—at issue. The particular kind of necessity at issue can be looked at from our two distinct perspectives of philosophical and natural relations. Hume explicitly recognizes this with respect to causal necessity: 'I define necessity in two ways, conformable to the two definitions of *cause*, of which it makes an essential part. I place it either in the constant union and conjunction of like objects, or in the inference of the mind from the one to the other' (2. 3. 2. 4). This must seem odd, since Hume has argued in 1. 3. 14 that necessity is purely in the mind; it does not even have an analog when considered in things themselves. It is surprising to see Hume turn around here and say that necessity *can* be considered as a relation between things. For in this sense, as a philosophical relation, causal necessity would seem to be nothing.

For ease of reference, let us call necessity as a philosophical relation NPR, and as a natural one, NNR. A helpful start will be to consider necessity in these two ways with regard to a relation besides causation. Let us take, then, mathematical equality. To say that $2 + 2 = 4$ is a relation characterized by NNR is to say that, once the mind forms the idea '$2 + 2$,' it is determined to form the idea of '4.' At the same time, however, one can also think of the relation of equality involved here without attending to the determination of the mind; thus, we treat it as NPR when we simply focus on the ideas themselves. That is, equality as falling under NNR allows us to think of the items involved and neglect the activity of the mind in moving between them, and hence as falling under NPR.

What does Hume mean, then, when he says that the mind 'discovers' relations? Given the distinction between philosophical and natural, we can say that the mind, having formed a natural relation, can then detect a recurrence of the same circumstances as initially prompted the transition from one idea to another. That is, we can discover that philosophical relations obtain in the sense that we can discover new perceptions that prompt the imagination to go to work in the same way as the old ones. We can discover necessity in the sense that we can flag sets of perceptions sufficiently resembling those we have already connected in imagination.

Hume thus modifies conceptual foundationalism; a better name, if one is needed, might be 'psychological foundationalism': the association of perceptions in the mind, via the activity of comparison that results in complex ideas of relations, is due to the imagination, and not to the intrinsic natures of the ideas compared. This alone lets us explain why Hume says that 'reason is nothing but a wonderful and unintelligible instinct in our souls' (1. 3. 16. 9).[9]

---

[9] See also *T* 1. 3. 8. 12: ''Tis not solely in poetry and music, we must follow our taste and sentiment, but likewise in philosophy. When I am convinc'd of any principle, 'tis only an idea, which strikes me more strongly.' For a broadly similar account of Hume's take on reason, see Owen (1999).

27

The Definition of Causation

Some relations, then, have a dual nature: we can think of them by thinking of the relata, or by reflecting on the decisive tendency of the mind to join the relata in thought. It is no surprise, then, that Hume offers two definitions of causation in the *Treatise*, one suited to each conception.[1] Many commentators, however, have indeed found it to be a surprise, and at least prima facie, a disappointment. For the two definitions are not coextensive. And yet Hume does not say that he is simply offering informal characterizations; instead, he claims he will unite all of his preceding reasoning 'to form an exact definition of the relation of cause and effect' (1. 3. 14. 30).

27.1 THE PROBLEM

Hume writes,

We may define a CAUSE to be 'An object, precedent and contiguous to another, and where all the objects resembling the former are plac'd in like relations of precedency and contiguity to those objects, that resemble the latter.' If this definition be esteem'd defective, because drawn from objects foreign to the cause, we may substitute this other definition its place, namely, 'A CAUSE is an object precedent and contiguous to another, and so united with it, that the idea of the one determines the mind to form the idea of the other, and the impression of the one to form a more lively idea of the other.' (1. 3. 14. 31)[2]

While these definitions are supposed to define one and the same concept—causation—they do not include the same objects in their extension. For

---

[1] In fact, Hume offers five definitions: two in the *Treatise*, and three in the *Enquiry*. I shall not attend to the differences between the pairs of definitions in the *Treatise* and the *Enquiry*; I think they are equivalent. The fifth definition, or, if one prefers, characterization, which comes in the *Enquiry*, is, to modern ears, deeply misleading: Hume glosses the first definition by saying, 'where, if the first object had not been, the second never had existed' (*E* 7. 29). This sounds as if Hume thinks constant conjunction can support counterfactuals. Happily, Anne Jaap Jacobson (1986) has shown that this is not Hume's meaning.

[2] Perhaps this section would be better titled '*A* problem,' since Hume's second definition, what I call C2 below, seems to be circular. Hume says that one perception 'determines the mind' to form another; what is this determination, other than causation? I think the best answer here is that C2 presupposes C1: this determination itself is merely a constant conjunction.

example, a world without minds would have causes according to the first (call it 'C1'), but not the second ('C2').[3] Similarly, an unnoticed constant conjunction counts as a cause under C1, but not C2. An increasing number of complicated and often ingenious solutions have been devised, the most plausible of which, Don Garrett's, appeals to an ideal observer.[4] I think the answer can be found closer to home.

Note that Hume's first definition of cause is the same as his first definition of necessity, i.e., necessity as a philosophical relation (NPR). This is confusing, since necessity is supposed to be a constituent of the idea of cause and effect, not one and the same thing. The confusion is inevitable, since Hume takes causation and necessity to be both philosophical and natural relations, and yet he makes necessity in both senses an essential part of the former. None of this need obscure his view, however.

Given the fundamental status of natural relations, the first definition of cause, that is, of causality as a philosophical relation, is bound to seem odd, just as the first definition of necessity did. It omits the very thing Hume set out to look for as he wandered through the neighboring fields. The oddity disappears once we recall that constant conjunction is the source or prompt for our impression of necessity. And like all relations with a dual nature, we can think of causation as involving a determination of the mind or not.

This fits well with how Hume presents his definitions. He says that the only difference between them is that they 'present . . . a different view of the same object' (1. 3. 14. 31). Note that he says 'object,' not 'objects': the different view is of the same relation, now as philosophical, now as natural, not of the relata.

Given the systematic ambiguity Hume sees in words for relations that have a dual nature,[5] there is no reason to expect that he intends his definitions to be coextensive, even though they are definitions of one and the same thing, namely, causation. Natural and philosophical relations have different extensions, as we have already seen. If we take resemblance as a natural relation, a tsar's undergarments and Fess Parker's teeth bear no resemblance to each other at all. If we take it as a philosophical relation, then they do: they resemble each other in so far as both are objects, are both yellow, etc.

Thus, I think it is a mistake to try to make Hume's definitions coextensive. Given this relation's dual nature as captured in those definitions, there is no reason to think C1 and C2 will necessarily include all and only the same items,

---

[3] This way of putting matters is not quite satisfactory, since, if objects are just perceptions, there is no such thing as a world of objects that does not also contain minds. But a sufficiently similar scenario can be created by imagining a world in which the perceptions are not common to a single mind but rather spread out among a great many, with no two perceptions constantly conjoined in the experience of a single mind.

[4] See Richards (1965) and Garrett (1997). For more on Garrett, see the following section.

[5] See *T* 1. 1. 5. 1, discussed above.

any more than we would expect resemblance or contiguity to have identical extensions when taken in these two quite different senses.

In fact, it is vital to Hume's psychological project that the two definitions *not* be coextensive and that there be a legitimate sense in which mere constant conjunction suffices for causation. For any causal sequence $a-b$, and any subject S, there will be a time at which S does not feel any determination of the mind to think of b on perceiving a. Upon repeated exposure to the constant conjunction of a and b, however, S does come to feel it. How then will S regard the previous observations of $a-b$ sequences? Surely S will see them as causally connected, even though he did not at the time experience any impulse to transfer his thoughts from a to b. This is not to be accounted for by a kind of projection from his current state into his past: he is realizing that a and b, all along, matched his idea of causation, and he is quite right about this. Only causation *qua* philosophical relation can allow us to think in this way.

27.2 SUBJECTIVISM OR PROJECTIVISM?

'Watching Charlie Rose last night put me to sleep.'

'If only I hadn't turned on PBS, I wouldn't have fallen asleep at 9.'

Claims such as these seem to assert something that goes well beyond the mere sequence of events. And yet, given Hume's first definition of 'cause,' this is all they do mean. It certainly feels as if there's a big difference between our first claim and 'Last night I watched Charlie Rose and then went to sleep.' But if Hume is right, the difference is illusory. What is more alarming, when causal claims do assert something more than mere regularity, as is explicit in the second, counterfactual claim, they assert something about our own psychology, namely, that one's mind is determined to form one idea on the basis of another. Any such claim then belongs partly to the realm of autobiography. But the assertions above at least on their surface purport to be about how events are in fact connected, not how I happen to feel about them.

There are in fact three distinct problems lurking here. We must keep in mind that the supposition of Causation is *not* an inevitable result of the workings of the imagination. The mind's tendency to 'spread itself' on the world can and should be resisted if we are to avoid philosophical bad faith. The person in the street, I shall argue, is not projecting anything onto the world; this is a mistake only a philosopher could make. By contrast, a belief in the external world (understood as the continuity of perceptions) is unavoidable, at least in everyday life.

Our first problem, then, is to give a diagnosis of the pre-theoretical tendency to think of distinct perceptions as necessarily connected. Second, we must find a way to account for the Causal realist's claims; what is the cognitive content of their assertions, if, as I have argued, the concept of Causation is incoherent?

Finally, given Hume's own definitions of 'cause,' how can he avoid the prima facie repugnant view that causal claims are partly self-reports?

These questions are not as independent as they seem. Barry Stroud has forcefully argued that Hume must be able to give an account of the Causal realist's claims as coherent (though unjustified or false), on pain of collapsing into subjectivism.[6] I shall argue that the situation is not nearly so dire. We can take the sting out of the subjectivist element in Hume's views, both of his antagonists' and of his own causal talk, by recalling the basic features of his philosophy of language.

As I have put them, our problems correspond to the three stages of development Hume isolates in his sketch of the history of causation: that of the vulgar philosophy, the false, and the true (1. 4. 3. 9–10).[7] Each stage is a natural development in one's mental life that has been recapitulated, however haphazardly, in the philosophical world.[8] The 'vulgar' philosophy is a pre-theoretical view whose force is felt in ancient philosophy, where Hume finds the unreflective claim that the objects we experience are linked by logical necessity. Since custom makes it hard for us to separate ideas, however distinct, the vulgar naturally come to believe that there is a connection between distinct existences. The 'false philosophy' arises when philosophers get a glimmer of the truth: they discern that the objects that we experience, at least, are not logically necessarily connected. But infected with the aboriginal error of the vulgar philosophy, they avoid the conclusion that no objects are necessarily connected by attributing causal power to unobservable agents. Occult qualities or faculties soothe the conscience of the learned by hiding their ignorance. These thinkers are astute enough to see the error of the vulgar, but instead of rejecting Causation, they squabble among themselves. It is not surprising that their barbs should find ready victims, since they are one and all in pursuit of an illusory goal.

Let us begin with the first stage. Those who project necessity onto the world are not the average person in the street, any more than the defender of the doctrine of double existence is the average believer in the external world. Put most simply, at the earliest stages of our individual and collective development, we think there's a logically necessary connection between distinct objects because

---

[6] Stroud writes, 'if necessity just *is* a determination of the mind, then that is what our idea of necessity is an idea *of*. But if our idea of necessity is an idea of a determination of the mind, then in ascribing necessity to the connections between things we are simply saying something about our own minds. We are saying that our minds do, or would, expect a thing of one kind after having observed a thing of another kind. This would commit Hume to the subjectivistic or psychologistic view that every causal statement we make, whatever its putative subject matter, is at least partly a statement about us. Rather than expressing a belief that something is objectively true of the connection between two objects or events, we would merely be asserting that something is happening or will happen in our minds when we observe certain objects or events' (1977: 83).

[7] Note that this tripartite model mirrors the development of thought about the external world.

[8] See *T* 1. 4. 3. 9: 'we may observe a gradation of three opinions, that rise above each other, as the persons, who form them, acquire new degrees of reason and knowledge.'

we cannot see the difference between them. Habit makes it difficult to separate the idea of, say, fire from that of heat.

If I don't see how fire and heat can be separated, at least in thought, I'll think they're necessarily connected, simply because I'll run the two together in mind. Summarizing this first stage, Hume writes,

'Tis natural for men, in their common and careless way of thinking, to imagine they perceive a connexion betwixt such objects as they have constantly found united together; and because custom has render'd it difficult to separate the ideas, they are apt to fancy such a separation to be in itself impossible and absurd.   (1. 4. 3. 9)

This is not the outcome of any sophisticated philosophical reasoning. It is simply the uncorrected brute force of custom as it were squashing distinct ideas together. It then takes mental effort to pull them apart. In this way, the vulgar naturally come to believe that there is a necessary connection between distinct existences.

Thus, the nonsense-speakers and self-contradictors are not ordinary folk but philosophers. (This is not to say, of course, that the ordinary folk are *right*—they are making a natural error, though an error nonetheless.) The unreflective view does not project an internal impression of necessity onto things that cannot, by their nature, have it. Instead, the mind, having experienced the constant conjunction of *a* and *b*, *blurs the line* between *a* and *b* and confounds these quite distinct perceptions. So we do not need to appeal to projection to account for our impulse to believe that cause and effect are necessarily connected, simply because most of the time we believe a cause and its effect are one and the same thing. Their seeming inseparability is a result of our not seeing their distinctness, not of our projecting an idea of necessity onto them.

When philosophers, presumably under the influence of this habit, come to formulate a doctrine of causation, they do not make quite the same mistake. Instead, Hume thinks, they *project* an idea of necessity into a new context where it does not belong.

What is this new mistake? According to Barry Stroud, what happens here is that the mind falsely projects onto the world an idea it gets from reflecting on itself. Someone who insists on Causation is making an objective claim, albeit one that is always false. On his view, then, Causation is an intelligible, though ultimately mistaken, belief *about* the physical world.[9] The alternative, Stroud thinks, is subjectivism, according to which even Causal realists are ultimately doing nothing more than making self-reports: all one can mean by '*a* and *b* are *really, mind-independently, even if no one were here to think about them, from the God's-eye point of view*, Causally connected' is that 'I feel a compulsion to move from the idea of *a* to the idea of *b*.' This, Stroud thinks, makes a shambles of our ordinary linguistic practices.

[9] Stroud (1977: 86).

Unfortunately, the coherent projection view is doomed. For it to work, one must have an idea of necessary connection that is intelligibly applied to objects. The problem is that, our idea of necessity being what it is, to apply it in this way is to make a category mistake.[10] Consider Hume's diagnosis of the 'spreading' tendency, common to Malebranche's 'pagans':

'Tis a common observation, that the mind has a great propensity to spread itself on external objects, and to conjoin with them any internal impressions, which they occasion, and which always make their appearance at the same time that these objects discover themselves to the senses. Thus as certain sounds and smells are always found to attend certain visible objects, we naturally imagine a conjunction, even in place, betwixt the objects and qualities, tho' the qualities *be of such a nature as to admit of no such conjunction*, and really exist no where.   (1. 3. 14. 25; my emphasis)

Thus, for example, we are tempted to think that the taste of a fig has a spatial location (1. 4. 5. 11). The taste 'is suppos'd to lie in the very visible body.' But tastes are not the right kind of thing to have a location; they exist nowhere, and do not even 'admit' of place. This does not stop us from making the supposition. This is not because it is an intelligible, though necessarily false, belief; it is because one can suppose absolutely anything.[11] To say that x has a power to φ, in the realist's sense, is like saying that a sensation of taste exists in an object, and, most of the time, even the realist cannot be this confused.

Recall that when we ask after the 'ultimate and operating principle, as something, which resides in the external object, we either contradict ourselves or talk without meaning' (1. 4. 7. 5).[12] Discovering the true nature of necessity as a mere determination of the mind 'prevents our wishes' for an ultimate principle because it exposes them as incoherent. We can call this view, in opposition to Stroud's, 'incoherent projection.'

Part of Stroud's problem remains, however: if Hume is right, those who use causal language properly, according to Hume's own definition C2, will sometimes be making a self-report instead of a claim about the way the world is.

We can develop a plausible response by looking at a solution drawn from Don Garrett's contributions to the problem we have just discussed: the fact that C1 and C2 are not coextensive. On Garrett's view, we need not assume that a causal claim in sense C2 is a self-report. When Hume says that *the mind* is determined to form one perception on the basis of another, this phrase need not, Garrett thinks, be read as any particular mind, or the speaker's own. Instead, Hume might have been thinking of the mind of an ideally situated observer. A causal

---

[10] As Stanford (2002) persuasively argues.

[11] See Winkler (1991) on 18th-century uses of 'suppose.'

[12] Note that, as Winkler shows, the operative notion of contradiction is not the formal contemporary notion but the mentalistic notion of contradiction as intuitively introspectible repugnance or incompatibility. So my talk of incoherence and Hume's talk of self-contradiction need not be incompatible.

claim thus construed is obviously something one can be mistaken about and is not in any interesting sense a self-report.[13]

There is much that is appealing in this Garrett-inspired answer. Unfortunately, I think Garrett is wrong. As I have argued above, it is vital to Hume's project that C1 and C2 *not* turn out to be coextensive. In addition to the rather strained reading Garrett must give to Hume's use of 'the mind,' we have the further problem that it seems only to push Stroud's worry up a level. If one is not happy asserting something partly about her own mental states when she says 'Fire causes heat,' she should be no happier to be told that the real content of her claim concerns the counterfactual responses of an idealized mind. The gap between asserting something about the mind and asserting something about the things in question remains. I think there's a better answer in the offing.

That better answer requires that we recall Hume's philosophy of language. Consider how Stroud frames the objection: if we lack an idea of necessity that could sensibly be applied to objects, then 'every statement we make... is at least partly a statement about us. Rather than expressing a belief that something is objectively true of the connection between two objects or events, we would merely be asserting that something is happening or will happen in our minds when we observe certain objects or events.'[14] If Hume is right, when we speak correctly, we are in part asserting something about our own minds.

But if signification is indication, then every statement we make, whatever its content, is in a sense an assertion about our own minds. We are indicating an idea in our own minds and simultaneously helping to cause that idea to form in others. So there is nothing special about causal language per se. 'The mind' in C2 can really just be, as it seems to be, the speaker's mind, and not that of an imaginary observer. This might make a shambles of our linguistic practices, as Stroud suggests, but it is Hume's shambles.

A different way to put this point would be to say that our worry about the cognitive content of causal assertions has no real place in Hume's framework. It has to be translated, and in the process substantially altered, to affect him. A rough version of the question might be this: what idea is being signified by someone who says that watching Charlie Rose put him to sleep? In other words, what is the speaker indicating? The answer must be, an idea of these two events plus an indication that the speaker's mind is determined to move from one to the other. Both question and answer are considerably different from their contemporary counterparts.

To say that *a* causes *b* on C2, then, is not to *assert* something about one's own mind but to *indicate* how one associates the perceptions of *a* and *b*. This says nothing at all about why one associates them in this way. Now, it turns out

---

[13] Garrett in fact thinks that both C1 and C2 admit of 'subjective' and 'objective' readings. As long as C1 and C2 are read as both subjective or both objective, the definitions come out coextensive.   [14] Stroud (1977: 83).

that there's nothing more to *be* said by way of justification. But this isn't part of what a causal claim à la C2 indicates. Just as one who says 'The recreational torture of infants is wrong' is indicating an attitude toward the practice without thereby taking any meta-ethical stance at all, so one who says 'Fire causes heat' is indicating an attitude toward those two objects or events without offering anything further. This can be the case even when that attitude exhausts the things one could sensefully indicate about those events.[15]

---

[15] Something must be said here about the normative force of causal claims; after all, Hume gives us 'rules by which to judge of cause and effect' at *T* 1. 3. 15. Owen (1999: 206) argues that, for Hume, causal reasoning is analogous to augury: there are clearly delineated norms within the practice of augury, even though they have no further justification. Morris (2005: 90) points out a disanalogy: unlike augury, causal reasoning is not a practice we can step back from, as it were, and evaluate from a further standpoint. After all, augury is not in fact successful as a means of prediction, and this last is a judgment we make by holding augury to the standards of causal inference. The view implied by my arguments in this section is consistent with Morris's own; Hume's claim that reason is a 'wonderful and unintelligible instinct' applies to causal reasoning as much as to any other form.

Conclusion

Conclusions are a standing temptation for authors. One either tediously sums up all that has gone before or indulges in remarks which the work does not strictly justify. I have chosen to succumb to each in turn.

Consider one of the questions with which we began: Why does the conception of causal necessity as *logical* necessity so outlive the notion of powers on which it was based? In one way, the question itself is wrong. For the Aristotelian conception of power was not discarded so much as reinvented during the modern period, issuing in distinct models of causation. It is a mistake to think of the scholastic concept of power as lingering on without justification, long after it was unmoored by the 'new' philosophy. Instead, it was adopted and transformed (or, if you prefer, transmogrified).

Confronted with the challenge of ontological mechanism, the Aristotelian view splits off in two directions. While both preserve the notion that causation requires logical necessitation, they take core features of the Aristotelian position to point toward different paths. On the Cartesian view expressed most clearly in Malebranche, genuine causal power requires the kind of directedness or *esse-ad* only minds can possess. In a world denuded of Aristotelian forms, the only plausible locus for causal power is the mind, whether God's or our own. This is what I have called the 'cognitive' model: a cause is directed at its effect in virtue of the effect's being included in that cause's intentional content. Coupled with the intentionality requirement, the new ontology results in a top-down view.

But the 'bottom-up' feature of Aristotelianism also retains its influence, and leads philosophers like Régis, Boyle (on occasion), and most clearly Locke to try to remake Aristotelian powers in the mechanist image. The struggle is to reconcile ontological and course-of-nature mechanism by crafting a picture of powers that can fit in a world exhaustively characterized by mechanical qualities. To do this, Locke reduces powers to relations, and relations to their relata. The intrinsic mechanical qualities of bodies are necessarily connected to their effects; no supernatural agency is required to immediately generate the phenomena we see around us. Trying to preserve the strong bottom-up impulses of Aristotelianism while at the same time remaining faithful to the new ontology, these philosophers produce what I have called the 'geometrical' model of causation.

In another way, the question why the moderns as a whole take causation to be logical necessitation is legitimate: why, given the obvious alternative of causation as *nomo*logical necessitation, does this conception persist? The answer is that this alternative is in no way obvious: it is neither epistemically available to the philosophers we have looked at, nor transparently correct. What follows is devoted to supporting this last remark.

I have argued that the concept of law, deployed in the new context of physics, retains some features of its ancestor. As Boyle and Cudworth pointedly argue, a law that governs events needs an enforcer. Descartes and Malebranche, I have claimed, are both happy with this result. For these figures, any appeal to laws entails occasionalism. Neither of them suggests that the laws might operate 'on their own,' an alternative regarded on all hands as absurd. Even Newton, in his letter to Bentley, still thinks of laws as requiring an agent to implement them at every moment. The notion that laws might themselves necessitate events is seen, on the rare occasions that it *is* seen, as incoherent.

Among the lessons one might draw from the details of the modern debate as I have set it out is just how much turns on the questions of intentionality and meaning. Most recently, we have seen Hume's impoverished empiricism, which lacks the resources of Lockean definite description and hence makes causal realism unintelligible. And in examining possible responses to the 'no necessary connection' argument set forth by Malebranche and Hume, we saw just how indebted one's modal intuitions are to one's conception of intentionality and ontology generally. Malebranche and Hume find no difficulty whatever in conceiving of alternative courses of nature; any sublunary event might be followed by any other. It is open to their opponents to argue that they have either misdescribed what they are conceiving or failed to conceive anything at all. The starkest case would be the Aristotelian: given the presence in the mind of an intelligible species, itself drawn from the embodied substantial form, one cannot in fact conceive of fire that failed to burn paper. It is only inattention to these forms, or an incomplete grasp of them, that makes such events *seem* coherently conceivable and hence possible. Similar responses can be generated, as we have seen, on behalf of Régis and Locke.

This skepticism about the power of conceivability arguments brings us to the question of laws of nature in the contemporary scene. Hardly anyone finds the top-down picture of Descartes and Malebranche attractive these days. But the top-down view lives on in current debates. The notion that laws of nature are facts that obtain entirely independently of the non-nomic facts is very much alive, even though the occasionalism the moderns think is required by such a picture is not. And such a picture has derived considerable support from appeals to modal intuition.

Consider two worlds, w_1 and w_2, each of which contains nothing but X-particles and Y-fields. In these worlds, as it happens, no X-particle ever enters a Y-field. Nevertheless, it might be a law in w_1 that any X-particle entering a Y-field acquires spin-up, while in w_2, any such particle acquires spin-down. So here we have two worlds identical in all their non-nomic respects differing in the laws of nature that obtain in them.[1] Note that this kind of thought experiment, if successful, threatens not just 'Humean' supervenience theories but also bottom-up conceptions, according to which events are necessarily connected in virtue of the intrinsic qualities of the objects that figure in them.

Compare a parallel argument for the non-reducibility of dispositions. Take two qualitatively identical worlds, one of which includes object *a*, and the other, object *b*, neither of which encounters situation *x*. Now just imagine *a* behaving one way in situation *x* and *b* in another way. Presto! Dispositional facts do not supervene on non-dispositional facts. If this has the air of getting something for nothing, so does the nomological thought experiment.

For my part, I find it hard to imagine that the counterfactuals the experiment appeals to could be the result of nothing more than the laws *tout court*. If there really is no non-nomic difference between the two worlds, I cannot see how different counterfactuals could obtain in them. In fact, I draw the opposite lesson from this thought experiment: supposing that laws of nature swing free, in this way, from the objects that 'obey' them is a deeply *counter*-intuitive view.[2] My opponents can of course reply that my inability to find the situation possible is simply a reflection of my own biases. But this goes both ways. As in so many other cases, thought experiments can illustrate views and their consequences, but they do not provide genuine arguments for them.

And now we come to the *least* justified point I shall allow myself to make here: I think the very notion of a law of nature, and all the attendant talk of nomological necessity, is either vacuous or incoherent. Laws of nature are either summaries of regularities, in which case they supervene on the non-nomic facts, or they are . . . what, exactly? Not features of God's will, presumably. If they were, then we could make sense of laws of nature, though in a way that is very hard to take seriously. This is no place for a survey of contemporary realist positions on laws of nature.[3] But I do think the inability of those moderns who buy into the top-down position to detach their views from theology is indicative not of a weakness of mind but of just the opposite: if an intelligible notion of laws in this sense requires a divine agent, then so be it.

Now, to say that nomic facts supervene on non-nomic facts is not to endorse contemporary or classical Humeanism. For among those non-nomic facts might be not just regularities but mind-independent features that explain and ground those regularities. The best candidates for such features are the intrinsic properties of bodies. This, of course, just is the bottom-up picture Régis and Locke struggled to preserve. As I have argued, abandoning nomological modality for logical is not

---

[1] Versions of this kind of thought experiment can be found in Michael Tooley (1977) and John Carroll (1990); my own statement follows Carroll's formulation. See Helen Beebee (2004) and Barry Loewer (2004) for discussion.

[2] My thoughts here have turned out to be very much in line with those of George Molnar. As Molnar puts it, 'it is causal dependence that explains any counterfactual dependence and increase in probability rather than vice versa' (2003: 187).   [3] See above, Ch. 1 n. 7.

as absurd as it at first seems.[4] On the view I am recommending, all talk of 'laws of nature' is a dispensable convenience. If we can purge ourselves of Humean intuitions on one hand and of the equally ill-founded intuitions that drive the top-down picture on the other, we will finally see our way past the miasma of confusion that surrounds the notion of a law of nature to a more sensible view: one that brings causation back down to earth. Where it belongs.

---

[4] It would be a mistake to think the view I am recommending cannot account for probabilistic causation. So modified, the view would hold that it is a matter of logical necessity that the presence of property F makes result *e* more or less probable (to whatever degree).

References

Adams, M. M. (1987), *William Ockham*, 2 vols (Notre Dame: University of Notre Dame Press).

——(2007), 'Powerless Causes: The Case of Sacramental Causality,' in P. Machamer and G. Wolters (eds), *Thinking about Causes: From Greek Philosophy to Modern Physics* (Pittsburgh: University of Pittsburgh Press).

Alexander, Peter (1985), *Ideas, Qualities, and Corpuscles* (Cambridge: Cambridge University Press).

Anscombe, G. E. M. (1993), 'Causality and Determination,' in E. Sosa and M. Tooley (eds), *Causation* (Oxford: Oxford University Press).

Anstey, Peter R. (2000), *The Philosophy of Robert Boyle* (London: Routledge).

Aquinas, St Thomas (1945), *Basic Writings*, ed. Anton Pegis, 2 vols (New York: Random House).

——(1993), *Selected Philosophical Writings*, ed. T. McDermott (Oxford: Oxford University Press).

Aristotle (1984), *The Complete Works of Aristotle*, ed. Jonathan Barnes, 2 vols (Princeton: Princeton University Press).

——(1989), *Prior Analytics*, trans. Robin Smith (Indianapolis: Hackett).

Armstrong, D. M. (1978), *Universals and Scientific Realism*, 2 vols (Cambridge: Cambridge University Press).

——(1983), *What Is a Law of Nature?* (Cambridge: Cambridge University Press).

——(1989), *Universals: An Opinionated Introduction* (Boulder, Colo.: Westview Press).

——(1999), 'Comment on Ellis,' in Howard Sankey (ed.), *Causation and Laws of Nature* (Dordrecht: Reidel).

——(2002), 'The Causal Theory of the Mind' (1981), in D. Chalmers (ed.), *Philosophy of Mind* (Oxford: Oxford University Press).

Ayer, A. J. (1946), *Language, Truth and Logic* (New York: Dover).

Ayers, M. R. (1981), 'Mechanism, Superaddition, and the Proof of God's Existence in Locke's *Essay*,' *Philosophical Review*, 90: 210–51.

——(1991), *Locke: Epistemology and Ontology*, 2 vols in 1 (London: Routledge).

——(1996), 'Natures and Laws from Descartes to Hume,' in G. A. J. Rogers and S. Tomaselli (eds), *The Philosophical Canon in the 17th and 18th Centuries* (Rochester, N.Y.: University of Rochester Press).

——(1998), 'The Foundations of Knowledge and the Logic of Substance: The Structure of Locke's General Philosophy,' in Vere Chappell (ed.), *Locke* (Oxford: Oxford University Press).

——and Daniel Garber (eds) (1998), *The Cambridge History of Seventeenth-Century Philosophy*, 2 vols (Cambridge: Cambridge University Press).

Bacon, Francis (2000), *The New Organon*, ed. L. Jardine and M. Silverthorne (Cambridge: Cambridge University Press).

Beauchamp, Tom, and Alexander Rosenberg (1981), *Hume and the Problem of Causation* (Oxford: Oxford University Press).

Beebee, Helen (2004), 'The Non-Governing Conception of Laws of Nature,' in John Carroll (ed.), *Readings on Laws of Nature* (Pittsburgh: University of Pittsburgh Press).
—— (2006), *Hume on Causation* (New York: Routledge).
Bennett, Jonathan (1971), *Locke, Berkeley, Hume: Central Themes* (Oxford: Clarendon Press).
Berkeley, George (1949–58), *The Works of George Berkeley*, ed. A. A. Luce and T. E. Jessop, 8 vols (London: Thomas Nelson).
Blackburn, Simon (1993), 'Hume and Thick Connexions,' in *Essays in Quasi-Realism* (Oxford: Oxford University Press).
Boyle, Robert (1772), *The Works of the Honourable Robert Boyle*, 6 vols (London: T. Birch).
—— (1991), *Selected Philosophical Papers of Robert Boyle*, ed. M. A. Stewart (Indianapolis: Hackett).
—— (1999), *The Works of Robert Boyle*, ed. M. Hunter and E. Davis, 14 vols (London: Pickering and Chatto).
Brandt, Frithiof (1927), *Thomas Hobbes's Mechanical Conception of Nature* (Copenhagen: Levin and Munksgaard).
Broughton, Janet (1987), 'Hume's Ideas about Necessary Connection,' *Hume Studies*, 13: 217–44.
Carraud, Vincent (2002), *Causa sive Ratio: La Raison de la cause, de Suarez à Leibniz* (Paris: Presses Universitaires de France).
Carroll, John (1990), 'The Humean Tradition,' *Philosophical Review*, 99: 185–219.
—— (ed.) (2004), *Readings on Laws of Nature* (Pittsburgh: University of Pittsburgh Press).
Cartwright, Nancy (1980), 'Do the Laws of Physics State the Facts?' *Pacific Philosophical Quarterly*, 61: 75–84.
Chambers, Ephraim (1728), *Cyclopaedia, or, An Universal Dictionary of Arts and Sciences* (London).
Charleton, Walter (1654), *Physiologia Epicuro-Gassendo-Charltoniana, or, A Fabrick of Science Natural, upon the Hypothesis of Atoms Founded by Epicurus* (London: Thos Heath).
Clarke, Desmond (1989), *Occult Powers and Hypotheses* (Oxford: Clarendon Press).
Clatterbaugh, Kenneth (1995), 'Cartesian Causality, Explanation, and Divine Concurrence,' *History of Philosophy Quarterly*, 12: 195–207.
—— (1999), *The Causation Debate in Modern Philosophy* (New York: Routledge).
Cohen, L. Bernard, and George E. Smith (eds) (2002), *The Cambridge Companion to Newton* (Cambridge: Cambridge University Press).
Cotes, Roger (2004), Preface to the Second Edition, in Isaac Newton, *Philosophical Writings*, ed. Andrew Janiak (Cambridge: Cambridge University).
Coventry, Angela (2003), 'Locke, Hume, and the Idea of Causal Power,' *Locke Studies*, 3: 93–112.
Cudworth, Ralph (1837), *The True Intellectual System of the Universe*, 2 vols (New York: Gould and Newman).
Curley, Edwin (1972), 'Primary and Secondary Qualities,' *Philosophical Review*, 81/4: 438–64.
Des Chene, Dennis (1996), *Physiologia* (Ithaca: Cornell University Press).

—— (2006), 'From Natural Philosophy to Natural Science,' in D. Rutherford (ed.), *The Cambridge Companion to Early Modern Philosophy* (Cambridge: Cambridge University Press).

Digby, Sir Kenelm (1657), *Two Treatises: In the one of which, the nature of bodies, in the other the nature of mans soule is looked into: in way of discovery of the immortality of reasonable soules* (London).

Disalle, Robert (2002), 'Newton's Philosophical Analysis of Space and Time,' in L. B. Cohen and George E. Smith (eds), *The Cambridge Companion to Newton* (Cambridge: Cambridge University Press).

Downing, Lisa (1972), 'Locke, Boyle, and the Distinction Between Primary and Secondary Qualities,' *Philosophical Review*, 81/4: 438–69.

—— (1998), 'The Status of Mechanism in Locke's *Essay*,' *Philosophical Review*, 107/3: 381–414.

—— (2007), 'Locke's Ontology,' in Lex Newman (ed.), *The Cambridge Companion to Locke's 'Essay'* (Cambridge: Cambridge University Press).

Dretske, Fred (1977), 'Laws of Nature,' *Philosophy of Science*, 44: 248–68.

Ellis, Brian (1999*a*), 'Causal Powers and Laws of Nature,' in Howard Sankey (ed.), *Causation and Laws of Nature* (Dordrecht: Reidel).

—— (1999*b*), 'Response to David Armstrong,' in Howard Sankey (ed.), *Causation and Laws of Nature* (Dordrecht: Reidel).

—— (2001), *Scientific Essentialism* (Cambridge: Cambridge University Press).

—— (2002), *The Philosophy of Nature* (Montreal: McGill-Queen's University Press).

Flage, Daniel (1981), 'Hume's Relative Ideas,' *Hume Studies*, 7/1: 55–73.

—— (1997), 'Hume's Missing Shade of Blue,' *Modern Schoolman*, 75: 55–63.

—— and Clarence Bonnen (1997), 'Descartes on Causation,' *Review of Metaphysics*, 50: 841–72.

Freddoso, Alfredo (1991), 'God's General Concurrence with Secondary Causes: Why Conservation Is Not Enough,' *Philosophical Perspectives*, 5: 553–85.

Gabbey, Alan (1980), 'Force and Inertia in the Seventeenth Century,' in Stephen Gaukroger (ed.), *Descartes: Philosophy, Mathematics, and Physics* (London: Harvester Press).

—— (2002), 'Newton, Active Powers, and the Mechanical Philosophy,' in L. B. Cohen and George E. Smith (eds), *The Cambridge Companion to Newton* (Cambridge: Cambridge University Press).

Garber, Daniel (1992), *Descartes' Metaphysical Physics* (Chicago: University of Chicago Press).

—— (1993), 'Descartes and Occasionalism,' in Steven Nadler (ed.), *Causation in Early Modern Philosophy* (University Park: Pennsylvania State University Press).

—— (2001), *Descartes Embodied* (Cambridge: Cambridge University Press).

Garrett, Don (1997), *Cognition and Commitment in Hume's Philosophy* (Oxford: Oxford University Press).

Gassendi, Pierre (1972), *The Selected Works of Pierre Gassendi*, ed. Craig Brush (New York: Johnson).

Gaukroger, Stephen (ed.) (1980), *Descartes: Philosophy, Mathematics, and Physics* (London: Harvester Press).

Geach, Peter (1980), *Reference and Generality* (Ithaca: Cornell University Press).

Glanvill, Joseph (1665), *Scepsis Scientifica, or, Confest Ignorance, the Way to Science; In an Essay of the Vanity of Dogmatizing, and Confident Opinion* (London: E. Cotes).

Goodman, Nelson (1983), *Fact, Fiction, and Forecast* (Cambridge, Mass.: Harvard University Press).

Gueroult, Martial (1980), 'The Metaphysics and Physics of Force in Descartes,' in Stephen Gaukroger (ed.), *Descartes: Philosophy, Mathematics, and Physics* (London: Harvester Press).

Hall, A. R. (1954), *The Scientific Revolution 1500–1800* (New York: Beacon Press).

Harré, Rom, and E. H. Madden (1975), *Causal Powers: A Theory of Natural Necessity* (Totowa, N.J.: Rowman and Littlefield).

Hatfield, Gary (1998), 'Force (God) in Descartes's Physics,' in John Cottingham (ed.), *Descartes* (Oxford: Oxford University Press).

Hattab, Helen (2001), 'The Problem of Secondary Causation in Descartes: A Reply to Des Chene,' *Perspectives on Science*, 8: 93–118.

—— (2007), 'Concurrence or Divergence? Reconciling Descartes's Physics with his Metaphysics,' *Journal of the History of Philosophy*, 45/1: 49–78.

Hausman, Alan (1967), 'Hume's Theory of Relations,' *Noûs*, 1/3: 255–83.

Henninger, Mark (1989), *Relations: Medieval Theories 1250–1325* (Oxford: Clarendon Press).

Hobbes, Thomas (1839–45), *The English Works of Thomas Hobbes*, ed. William Molesworth, 11 vols (London: Richards).

—— (1994), *Human Nature and De Corpore*, ed. J. C. A. Gaskin (Oxford: Oxford University Press).

Holdsworth, Richard (1648), *Directions for a Student in the Universitie*, repr. in H. F. Fletcher, *The Intellectual Development of John Milton* (Chicago: University of Illinois Press, 1961).

Hooker, Richard (1593), *Laws of Ecclesiastical Polity* (London).

Hutchison, Keith (1982), 'What Happened to Occult Qualities in the Scientific Revolution?' *Isis*, 73/2: 233–53.

Jackson, Reginald (1968), 'Locke's Distinction Between Primary and Secondary Qualities,' in C. B. Martin and D. M. Armstrong (eds), *Locke and Berkeley* (London: Macmillan).

Jacobson, Anne Jaap (1986), 'Causality and the Supposed Counterfactual Conditional in Hume's *Enquiry*,' *Analysis*, 46/3: 131–3.

Jacovides, Michael (2003), 'Locke's Construction of the Idea of Power,' *Studies in History and Philosophy of Science*, 34/2: 329–50.

Jesseph, Douglas (2005), 'Mechanism, Skepticism, and Witchcraft,' in T. M. Schmaltz (ed.), *Receptions of Descartes* (London: Routledge).

Jolley, Nicholas (1990*a*), 'Berkeley and Malebranche on Causation,' in J. Cover and M. Kulstad (eds), *Central Themes in Early Modern Philosophy* (Indianapolis: Hackett).

—— (1990*b*), *The Light of the Soul* (Oxford: Clarendon Press).

—— (2002), 'Occasionalism and Efficacious Laws in Malebranche,' in P. French and H. Wettstein (eds), *Renaissance and Early Modern Philosophy* (New York: Blackwell).

Kail, P. J. E. (2007), *Projection and Realism in Hume's Philosophy* (Oxford: Oxford University Press).

Kant, Immanuel (1958), *Critique of Pure Reason*, trans. N. K. Smith (London: Macmillan).

Keating, Laura (1993), 'Un-Locking Boyle: Boyle on Primary and Secondary Qualities,' *History of Philosophy Quarterly*, 10: 305–23.

Kemp Smith, Norman (1941), *The Philosophy of David Hume* (London: Macmillan).

Langton, Rae (2000), 'Locke's Relations and God's Good Pleasure,' *Proceedings of the Aristotelian Society*, 100: 75–91.

Lehoux, Daryn (2006), 'Laws of Nature and Natural Laws,' *Studies in History and Philosophy of Science*, 37: 527–49.

Leibniz, G. W. (1978), *Die Philosophischen Schriften von Gottfried Wilhelm Leibniz*, ed. C. J. Gerhardt (Hildesheim: Georg Olms).

——(1996), *New Essays on Human Understanding*, trans. and ed. Jonathan Bennett and Peter Remnant (Cambridge: Cambridge University Press).

Leijenhorst, Cees (2002), *The Mechanisation of Aristotelianism* (Leiden: Brill).

Lelevel, Henri de (1694), *La Vraye et la fausse metaphysique* (Rotterdam: Reiner Leers).

Lennon, Thomas (2007), 'The Eleatic Descartes,' *Journal of the History of Philosophy*, 45/1: 29–47.

Lewis, David (1983), 'New Work for a Theory of Universals,' *Australasian Journal of Philosophy*, 61: 343–77.

Livingston, Donald W. (1984), *Hume's Philosophy of Common Life* (Chicago: University of Chicago Press).

Locke, John (1823), *The Works of John Locke*, 10 vols (London: Tegg et al.).

——(1954), *Essays on the Law of Nature*, ed. W. von Leyden (Oxford: Clarendon Press).

Loewer, Barry (2004), 'Humean Supervenience,' in John Carroll (ed.), *Readings on Laws of Nature* (Pittsburgh: University of Pittsburgh Press).

McCann, Edwin (1994), 'Locke's Philosophy of Body,' in Vere Chappell (ed.), *The Cambridge Companion to Locke* (Cambridge: Cambridge University Press).

McCracken, Charles J. (1983), *Malebranche and British Philosophy* (Oxford: Clarendon Press).

Malebranche, Nicolas (1992), *Philosophical Selections*, ed. Steven Nadler (Indianapolis: Hackett).

Matthews, Michael R. (ed.) (1989), *The Scientific Background to Modern Philosophy* (Indianapolis: Hackett).

Milton, J. R. (1998), 'Laws of Nature,' in M. R. Ayers and Daniel Garber (eds), *The Cambridge History of Seventeenth-Century Philosophy*, 2 vols (Cambridge: Cambridge University Press), i.

Molnar, George (2003), *Powers* (Oxford: Oxford University Press).

More, Henry (1662), *A Collection of Several Philosophical Writings*, 2 vols (London: J. Flesher).

Morris, William Edward (2005), 'Belief, Probability, Normativity,' in Saul Traiger (ed.), *The Blackwell Guide to Hume's 'Treatise'* (London: Blackwell).

Mouy, Paul (1934), *Le Développement de la physique Cartesienne 1646–1712* (Paris: Vrin).

Mumford, Stephen (1998), *Dispositions* (Oxford: Oxford University Press).

Nadler, Steven (ed.) (1993a), *Causation in Early Modern Philosophy* (University Park: Pennsylvania State University Press).

——(1993b), 'Occasionalism and General Will in Malebranche,' *Journal of the History of Philosophy*, 31/1: 31–47.

Nadler, Steven (1996), '"No Necessary Connection": The Medieval Roots of the Occasionalist Roots of Hume,' *The Monist*, 79/3: 448–66.

—— (1998), 'Doctrines of Explanation in Late Scholasticism and in the Mechanical Philosophy,' in M. R. Ayers and Daniel Garber (eds), *The Cambridge History of Seventeenth-Century Philosophy*, 2 vols (Cambridge: Cambridge University Press), i.

—— (ed.) (2000), *The Cambridge Companion to Malebranche* (Cambridge: Cambridge University Press).

Newton, Isaac (1999), *The 'Principia': Mathematical Principles of Natural Philosophy*, trans. L. Bernard Cohen and Anne Whitman (Berkeley: University of California Press).

—— (2004), *Philosophical Writings*, ed. Andrew Janiak (Cambridge: Cambridge University Press).

Ockham, William of (1974), *Ockham's Theory of Terms: Part I of the Summa Logicae*, trans. M. Loux (Notre Dame, Ind.: University of Notre Dame Press).

Ott, W. (2002), 'Propositional Attitudes in Modern Philosophy,' *Dialogue*, 41/3: 551–68.

—— (2004a), 'The Cartesian Context of Berkeley's Attack on Abstraction,' *Pacific Philosophical Quarterly*, 85/4: 407–24.

—— (2004b), *Locke's Philosophy of Language* (Cambridge: Cambridge University Press).

—— (2006a), 'Descartes and Berkeley on Mind,' *British Journal for the History of Philosophy*, 14/3: 437–50.

—— (2006b), 'Hume on Meaning,' *Hume Studies*, 32/2: 233–52.

—— (2008a), 'Régis's Scholastic Mechanism,' *Studies in History and Philosophy of Science*, 39/1: 2–14.

—— (2008b), 'Causation, Intentionality, and the Case for Occasionalism,' *Archiv für Geschichte der Philosophie*, 90/2: 165–87.

Owen, David (1999), *Hume's Reason* (Oxford: Oxford University Press).

—— (forthcoming), 'Locke and Hume on Belief, Judgment, and Assent,' *Topoi*.

Pessin, Andrew (2003), 'Descartes's Nomic Concurrentism,' *Journal of the History of Philosophy*, 41/1: 25–49.

Putnam, Hilary (1973), 'Meaning and Reference,' *Journal of Philosophy*, 70/19: 699–711.

—— (1975), *Mind, Language, and Reality, Philosophical Papers*, ii (Cambridge: Cambridge University Press).

Radner, Daisie (1978), *Malebranche* (Amsterdam: Van Gorcum Assen).

Remnant, Peter (1979), 'Descartes: Body and Soul,' *Canadian Journal of Philosophy*, 9/3: 377–86.

Richards, Thomas J. (1965), 'Hume's Two Definitions of "Cause,"' *Philosophical Quarterly*, 15/60: 247–53.

Robison, Wade L. (1976), 'Hume's Ontological Commitments,' *Philosophical Quarterly*, 26/102: 39–47.

Rosenberg, Alexander (1993), 'Hume and the Philosophy of Science,' in D. F. Norton (ed.), *The Cambridge Companion to Hume* (Cambridge: Cambridge University Press).

Ruby, J. E. (1986), 'The Origins of Scientific "Law,"' *Journal of the History of Ideas*, 47/3: 341–59.

Sankey, Howard (ed.) (1999), *Causation and Laws of Nature* (Dordrecht: Reidel).

Schmaltz, Tad (1996), *Malebranche's Theory of the Soul* (Oxford: Oxford University Press).

—— (2003), 'Cartesian Causation: Body–Body Interaction, Motion, and Eternal Truths,' *Studies in History and Philosophy of Science*, 34: 737–62.

—— (2008), *Descartes on Causation* (Oxford: Oxford University Press).

Secada, Jorge (2000), *Cartesian Metaphysics* (Cambridge: Cambridge University Press).

Sergeant, John (1696), *The Method to Science* (London: W. Redmayne).

—— (1984), *Solid Philosophy Asserted Against the Fancies of the Ideists* (New York: Garland).

Shoemaker, Sydney (1980), 'Causality and Properties,' in Peter van Inwagen (ed.), *Time and Cause* (Dordrecht: Reidel).

Smith, George E. (2002), 'The Methodology of the *Principia*,' in L. B. Cohen and George E. Smith (eds), *The Cambridge Companion to Newton* (Cambridge: Cambridge University Press).

Sorabji, Richard (1980), *Necessity, Cause, and Blame* (Ithaca, N.Y.: Cornell University Press).

Stanford, P. Kyle (2002), 'The Manifest Connection: Causation and Meaning in Hume,' *Journal of the History of Philosophy*, 40/3: 339–60.

Stein, Howard (1990), 'On Locke, "The Great Huygenius, and the Incomparable Mr. Newton,"' in R. I. G. Hughes (ed.), *Philosophical Perspectives on Newtonian Science* (Cambridge, Mass.: MIT Press).

—— (2002), 'Newton's Metaphysics,' in L. B. Cohen and George E. Smith (eds), *The Cambridge Companion to Newton* (Cambridge: Cambridge University Press).

Steinle, Friedrich (2002), 'Negotiating Experiment, Reason, and Theology: The Concept of Laws of Nature in the Early Royal Society,' in W. Detel and C. Zittel (eds), *Wissensideale und Wissenkulturen in der frühen Neuzeit* (Berlin: Akademie-Verlag).

Stillingfleet, Edward (1697), *A Discourse in Vindication of the Doctrine of the Trinity: With an Answer to the Late Socinian Objections Against it from Scripture, Antiquity and Reason* (London).

Strawson, Galen (1989), *The Secret Connexion: Causation, Realism, and David Hume* (Oxford: Clarendon Press).

Stroud, Barry (1977), *Hume* (London: Routledge).

Stuart, Matthew (1998), 'Locke on Superaddition and Mechanism,' *British Journal for the History of Philosophy*, 6: 351–79.

Suárez, Francisco (1994), *On Efficient Causation*, trans. A. J. Freddoso (New Haven: Yale University Press).

—— (2002), *On Creation, Conservation, and Concurrence*, trans. A. J. Freddoso (South Bend, Ind.: St Augustine's Press).

Thalberg, Irving (1981), 'The Discovery of Nonsense,' in P. French and H. Wettstein (eds), *Midwest Studies in Philosophy*, 6 (Notre Dame, Ind.: University of Notre Dame Press).

Tooley, Michael (1977), 'The Nature of Law,' *Canadian Journal of Philosophy*, 7: 667–98.

Trentman, John A. (1982), 'Scholasticism in the Seventeenth Century,' in N. Kretzmann, A. Kenny, and J. Pinborg (eds), *The Cambridge History of Later Medieval Philosophy* (Cambridge: Cambridge University Press).

Watson, Richard (1993), 'Malebranche, Models, and Causation,' in Steven Nadler (ed.), *Causation in Early Modern Philosophy* (University Park: Pennsylvania State University Press).

Waxman, Wayne (1994), *Hume's Theory of Cognition* (Cambridge: Cambridge University Press).

Weinberg, Julius (1965), *Abstraction, Relation, and Induction* (Madison: University of Wisconsin Press).

Westfall, Richard (1978), *The Construction of Modern Science* (Cambridge: Cambridge University Press).

Wilson, Margaret (1979), 'Superadded Properties: The Limits of Mechanism in Locke,' *American Philosophical Quarterly*, 16: 143–50.

Winkler, Kenneth (1991), 'The New Hume,' *Philosophical Review*, 100/4: 541–79.

Woolhouse, Roger (1972), *Locke's Philosophy of Science and Knowledge* (New York: Barnes and Noble).

Wright, John P. (1983), *The Sceptical Realism of David Hume* (Minneapolis: University of Minnesota Press).

Index